Stem Cells - Laboratory and Clinical Research Series

ADULT STEM CELL SURVIVAL

STEM CELLS - LABORATORY AND CLINICAL RESEARCH SERIES

Focus on Stem Cell Research
Erik V. Greer (Editor)
2004. ISBN: 1-59454-043-8

Trends in Stem Cell Research
Erik V. Greer (Editor)
2005. ISBN: 1-59454-315-1

New Developments in Stem Cell Research
Erik V. Greer (Editor)
2006. ISBN: 1-59454-847-1

Neural Stem Cell Research
Erik V. Greer (Editor)
2006. ISBN: 1-59454-846-3

Stem Cell Therapy
Erik V. Greer (Editor)
2006. ISBN: 1-59454-848-X

Embryonic Stem Cell Research
Erik V. Greer (Editor)
2006. ISBN: 1-59454-849-8

Frontiers in Stem Cell Research
Julia M. Spanning (Editor)
2006. ISBN: 1-60021-294-8

Stem Cells and Cancer
Devon W. Parsons (Editor)
2007. ISBN: 1-60021-517-3

Hematopoietic Stem Cell Transplantation Research Advances
Karl B. Neumann (Editor)
2008. ISBN: 978-1-60456-042-8

Stem Cell Applications in Diseases
Mikkel L. Sorensen (Editor)
2008. ISBN: 978-1-60456-241-5

Stem Cell Applications in Diseases
Mikkel L. Sorensen (Editor)
2008. ISBN: 978-1-60876-925-4 (Online Book)

Leading-Edge Stem Cell Research
Prasad S. Koka (Editor)
2008. ISBN: 978-1-60456-268-2

Stem Cell Research Progress
Prasad S. Koka (Editor)
2008. ISBN: 978-1-60456-308-5

Stem Cell Research Progress
Prasad S. Koka (Editor)
2008. ISBN: 978-1-60876-924-7 (Online Book)

Progress in Stem Cell Applications
*Allen V. Faraday
and Jonathon T. Dyer (Editors)*
2008. ISBN: 978-1-60456-316-0

Developments in Stem Cell Research
Prasad S. Koka (Editor)
2008. ISBN: 978-1-60456-341-2

Developments in Stem Cell Research
Prasad S. Koka (Editor)
2008. ISBN: 978-1-60741-213-7 (Online Book)

Gut Stem Cells: Multipotent, Clonogenic and the Origin of Gastrointestinal Cancer
Shigeki Bamba and William R. Otto
2008. ISBN: 978-1-60456-968-1

Stem Cell Transplantation, Tissue Engineering and Cancer Applications
Bernard N. Kennedy (Editor)
2008. ISBN: 978-1-60692-107-4

Stem Cells
Philippe Taupin
2009. ISBN: 978-1-60692-214-9

Stem Cells
Philippe Taupin
2009. ISBN: ISBN: 978-1-61668-577-5 (Online Book)

Stem Cell Plasticity
Suraksha Agrawal, Piyush Tripathi and Sita Naik
2009. ISBN: 978-1-60741-473-5

Neural Stem Cells and Cellular Therapy
Philippe Taupin
2009. ISBN: 978-1-60876-017-6

Neural Stem Cells and Cellular Therapy
Philippe Taupin
2009. ISBN: 978-1-61668-660-4 (Online Book)

Adult Stem Cells Survival
Anatoly Konoplyannikov, Sergey Proskuryakov, and Mikhail Konoplyannikov
2010. ISBN: 978-1-61668-035-0

Pluripotent Stem Cells
Derek W. Rosales and Quentin N. Mullen
2010. ISBN: 978-1-60876-738-0

Stem Cells - Laboratory and Clinical Research Series

ADULT STEM CELL SURVIVAL

ANATOLY KONOPLYANNIKOV,
SERGEY PROSKURYAKOV
AND
MIKHAIL KONOPLYANNIKOV

Nova Science Publishers, Inc.
New York

Copyright © 2010 by Nova Science Publishers, Inc.

All rights reserved. No part of this book may be reproduced, stored in a retrieval system or transmitted in any form or by any means: electronic, electrostatic, magnetic, tape, mechanical photocopying, recording or otherwise without the written permission of the Publisher.

For permission to use material from this book please contact us:
Telephone 631-231-7269; Fax 631-231-8175
Web Site: http://www.novapublishers.com

NOTICE TO THE READER

The Publisher has taken reasonable care in the preparation of this book, but makes no expressed or implied warranty of any kind and assumes no responsibility for any errors or omissions. No liability is assumed for incidental or consequential damages in connection with or arising out of information contained in this book. The Publisher shall not be liable for any special, consequential, or exemplary damages resulting, in whole or in part, from the readers' use of, or reliance upon, this material.

Independent verification should be sought for any data, advice or recommendations contained in this book. In addition, no responsibility is assumed by the publisher for any injury and/or damage to persons or property arising from any methods, products, instructions, ideas or otherwise contained in this publication.

This publication is designed to provide accurate and authoritative information with regard to the subject matter covered herein. It is sold with the clear understanding that the Publisher is not engaged in rendering legal or any other professional services. If legal or any other expert assistance is required, the services of a competent person should be sought. FROM A DECLARATION OF PARTICIPANTS JOINTLY ADOPTED BY A COMMITTEE OF THE AMERICAN BAR ASSOCIATION AND A COMMITTEE OF PUBLISHERS.

LIBRARY OF CONGRESS CATALOGING-IN-PUBLICATION DATA
Konopliannikov, A. G. (Anatolii Georgievich)
  Adult stem cell survival / Anatoly Konoplyannikov, Sergey Proskuryakov, Mikhail Konoplyannikov.
    p. cm.
  Includes index.
  ISBN 978-1-61668-035-0 (softcover)
  1. Stem cells. 2. Cell death. I. Proskuriakov, Sergei. II. Konopliannikov, Mikhail. III. Title.
  QH588.S83K66 2009
  616'.02774--dc22
                                                            200905264

Published by Nova Science Publishers, Inc. ✝ New York

CONTENTS

Preface		**xi**
Chapter 1	Adult Stem Cells and Cell Renewal Systems	**1**
Chapter 2	Radiobiology of Adult Stem Cells	**3**
Chapter 3	Hyperthermia and Adult Stem Cells	**19**
Chapter 4	Phenomenon of "Ischemia/Reperfusion" for Adult Stem Cells	**23**
References		**29**
Index		**49**

PREFACE

The authors have shown recently that for the two types of adult stem cells (hematopoietic stem cells and intestinal epithelium stem cells), the "ischemia/reperfusion" reaction can be developed in vivo. The damaging action of this reaction onto stem cells can be diminished by an injection of a source of NO-radicals into the animal's body during reperfusion, since NO-radicals are capable of decreasing the negative effect of radicals produced after oxygen access into the ischemic tissues. The authors believe that these data may be used in the development of new approaches for the protection of cell systems of organism renewal after the damaging action of various agents.

Basing on our own and literature data, we analyze phenomenological problems of the death of several adult stem cells types, including hemopoietic and mesenchymal stem cells of bone marrow, stem cells of intestinal epithelium and other epithelia. We discuss potential mechanisms of cell death in vivo and in vitro after the action of ionizing radiation, hyperthermia and in the conditions of the "ischemia/reperfusion" reaction development.

Although the phenomenology of adult stem cells death after the action of ionizing radiation has become classical, the molecular mechanisms of the death of these stem cells types are still being actively studied. It has been shown that after the action of ionizing radiation both the repair of induced damage and programmed cell death are realized in stem cells, the latter process going via several specific molecular pathways. The effects of many modifiers of radiation damage in adult stem cells are related to their participation in a complex combination of the processes of genetic damage repair and cell thanatogenesis.

When adult stem cells undergo hyperthermia, other "targets" are damaged than those in the case of ionizing radiation. However, a similar formal approach developed earlier for the ionizing radiation action can be applied to describe the dependence between the cell survival and the damaging agent dosage. At the same time, the action of hyperthermia is characterized by a number of phenomena with no analogues existing in the case of ionizing radiation, for example, a known phenomenon of thermotolerance resulted from the increased production of the heat shock proteins.

We have shown recently that for the two types of adult stem cells (hematopoietic stem cells and intestinal epithelium stem cells) the "ischemia/reperfusion" reaction can be developed in vivo. The damaging action of this reaction onto stem cells can be diminished by an injection of a source of NO-radicals into the animal's body during reperfusion, since NO-radicals are capable of decreasing the negative effect of radicals produced after oxygen access into the ischemic tissues. We believe that these data may be used in the development of new approaches for the protection of cell systems of organism renewal after the damaging action of various agents.

Recently, studies on the cancer stem cells progenitors produced by the action of some carcinogens onto animals have started. The simultaneous studies on the survival of adult stem cells and cancer stem cells progenitors after the damaging action of ionizing radiation and hyperthermia may give rise to the improvement in the existing methods of malignant tumors treatment.

Chapter 1

ADULT STEM CELLS AND CELL RENEWAL SYSTEMS

An adult organism consists of a big number of different cell types (approximately 250-300) which constitute various cell populations of all the tissues and organs [1-2]. These cells can be damaged upon the action of different physical, chemical and biological agents. From the viewpoint of radiobiology, considering ionizing radiation as a possible cause of lethal damage of any cells in an organism, all the cell populations of an adult organism can be divided into two big groups based on their organization and reaction to irradiation. The first group involves cells fast or slowly regenerating during postnatal life due to cell division, while the second group represents systems not regenerating during this period of life [3-4]. Fast regenerating systems include bone marrow, intestinal epithelium, spermatogonial epithelium etc.; vascular endothelium and fibroblasts regenerate slowly. The cells which almost never regenerate under normal conditions are populations of parenchymal cells of liver, kidneys and lungs, though their regeneration can be stimulated by special conditions [3, 5-6]. It has been currently understood that the key role in the physiological and regenerative restoration of adult tissues belongs to adult stem cells [2, 5-8]. This fact applies to both fast and slowly regenerating tissues and organs, and even such tissues as nerve and muscle, which currently have no methods of successful stimulation of cell proliferation and until recently have been considered as systems unable to regenerate [9-11]. In the last few years, new data appeared on the presence of stem cells and, hence, on the possibility of cell reproduction in such cell populations, to replace dying cells and provide repair after damage [9-11].

The systems of fast cell regeneration have been most extensively studied; they are usually designed by a so-called hierarchical type - cell hierarchy from stem cells to functional elements. A. Michalowsky [3] suggested to call their reaction onto irradiation "H-systems (or hierarchical cell population) reaction". Non-regenerating (or, in some cases, slowly regenerating) cell systems suggested to be called "F-systems" (flexible cell lineage) are less well understood. The latter consist mainly of a homogeneous population of functionally competent cells being in the G_o –phase of cell cycle. Although the cells of H- and F-systems are pretty much similar in their radiosensitivity (tested as the loss of ability to "infinitely" reproduce), they are significantly different in the dynamics and level of adult stem cells death after irradiation, and also in the picture of radiation damage development.

The systems of fast cell regeneration usually involve three or more compartments: 1) Adult stem cells compartment. Those are divided into primitive, pluripotent (capable of generating not one, but several cell lineages simultaneously), and committed stem cells (which are the initial component of a certain cell lineage); 2) Proliferative-increasing cell pool. These cells reproduce and differentiate into the elements of the following compartment (often a maturing cells pool is considered as a subpool within this pool); 3) Pool of functional cells. Adult stem cells form the basis for such cell renewal systems and usually are the only source of their repopulation after massive loss caused by irradiation, action of chemicals or, less common, action of other damaging agents. Adult stem cells are characterized by such two fundamentally important properties as capability of long-lasting self-renewal (it is sometimes referred to as an infinite proliferative potential) and the ability to produce cells going into differentiation, in doing so they can initiate various cell lineage. For example, in the hematopoietic system a pluripotent stem cell is a common progenitor cell for all hematopoietic cell lineages [12]; in the intestinal epithelium it is the only common progenitor cell of four cell types: columnar epithelium or enterocytes in the small intestine and colonocytes in the large intestine, mucosecreting or goblet cells, enteroendocrine cells and Paneth cells [13]. In the last years, two more properties of adult stem cells have been discovered – their plasticity (a possibility of reprogramming the following differentiation of adult stem cells) [14] and a possibility of producing various types of pluripotent stem cells from embryonic stem cells in vitro [15].

Chapter 2

RADIOBIOLOGY OF ADULT STEM CELLS

There exist different methods for revealing adult stem cells, first of those methods were developed by radiobiologists in the 60-70s of the last century [16-20]. Following irradiation, an acute deficit of survived stem cells is created and formation of the cell colonies is observed in the regions of their progeny proliferation. During 20 sequential divisions, one survived stem cell produces more than a million of cells-progeny. These new cells form a peculiar "colony" that may be morphologically revealed and recorded. When applying this method to studying survival of hematopoietic stem cells (HSC), the analysis of colonies formation in the spleen appeared to be possible not only for own cells survived the radiation (endogenous colony test), but also for bone marrow cells transplanted into a lethally irradiated organism (exogenous colony test), as shown in pioneer works by Till and McCulloch [16-18]. In the same decade, Withers and Elkind developed the methods of intestinal "microcolonies" and"macrocolonies", identical in their approach to endogenous colony test for HSC, for studying post-radiation survival of stem cells of murine intestinal epithelium [19-20]. Later on, methods applying primary and constant cell cultures were developed for human and animal hematopoietic stem cells (and some other types too) [21-22]. A population of non-hematopoietic stromal stem cells found in bone marrow [23-24] and later named mesenchymal stem cells (MSC) also rapidly became an object of radiobiological studies [24, 27-28]. Another population of adult stem cells - spermatogonial stem cells – had drawn attention of Withers et al [29] and other radiobiologists [30] who used the previously described method of radiation devastation of this cell renewal system in this case as well. Analogous works on peculiar equivalents of "exocolonies" of non-irradiated and irradiated thyroid epithelium stem cells and other tissue-derived progenitor cells upon transplantation into different places in the body have been substantially less developed [31-34].

Before discussing the general laws of the death of different stem cells upon the action of ionizing radiation, here we briefly refer to the methods of quantitative description of the dependencies between the radiation dose (radiobiology literature conventionally uses Latin symbol D for this quantity) and cell survival (S). A relatively small number of defined schemes are currently used in the modern radiobiology for the description of dose dependencies of various cells' survival in vivo and in vitro. One of the most widely used schemes is a single-hit multitarget model [35-37]. Its widespread popularity is due to the fact that it allows one to extract the parameters characterizing cell radiosensitivity from the graph of cell survival vs. radiation dose. The formal description of this model is given by the equation:

$$S = 1 - [(1 - \exp(-D/D_0))]^n \text{ or, identically, } S = 1 - [(1 - e^{(-D/D_0)})]^n \quad (1)$$

where S is the fraction of survived cells (survival) after irradiation in the dose D, D_0 and n are constants (parameters) called "mean cell lethal dose" and "extrapolation number", correspondingly. The dose-effect curves plotted using semi-logarithmic scale (with the linear X-scale of the dose and the logarithmic Y-scale of the survival) have sigmoidal shape with the initial smoothly bent region followed by an almost linear (i.e. exponential in linear coordinates) dependence. It is easy to verify that, at sufficiently high D values the equation (1) transforms as follows:

$$\ln S \cong \ln n - D/D_0 \quad (2)$$

or, using decimal logarithms,

$$\lg S \cong \lg n - D/D_0 \times \lg e \quad (3)$$

These equations (2 and 3) describe the straight line which the dose-effect plot approaches at higher irradiation doses. They describe only a part of the "dose- cell survival" curve to be experimentally estimated using the methods of stem cells survival similar to endogenous colonies method for HSC. The measurement of this curve slope allows estimation of the mean cell lethal dose D_0, which is numerically equal to the dose decreasing cell survival by a factor of e, for example, from 1 to 1/e (i.e., from 1 до 0,37, or from 100% to 37%; therefore, sometimes the parameter D_{37} is used instead of D_0). In the formal approach of the "target theory" [38] the mean cell lethal dose is an absorbed dose which, for the evenly distributed events of target inactivation, is sufficient for precisely one hitting of the sensitive cell "target" with 100% cells damage. However, due to random distribution of events of target inactivation, according to the Poisson statistics such events will happen only in 63% of cells, while 37% of cells will remain intact. The point of the intersection of the linear part of the curve with Y-axis (i.e. at the zero dose of radiation) gives a value of the logarithm of "extrapolation number" (in formal terms of the "target theory" it corresponds to the number of "targets" in a cell). Thus, having plotted the linear part of the curve based on experimental data, a researcher gets an opportunity to simultaneously estimate two parameters characterizing cells radiosensitivity. Often one more parameter is used - "width" of the initial shoulder region (or quasithreshold dose) D_q, which is numerically equal to the dose of radiation, cut on the X-axis by the linear part of the dose-effect curve at the level of 100% survival. This parameter is related to the other two ones via the following expression:

$$D_q = D_0 \times \ln n \qquad (4)$$

It is conventional that D_q may be used to quantitatively characterize cells' ability to repair radiation damage, and that increase in this parameter means enhancement in the irradiated cells ability to regenerate after radiation damage. It is especially related to the so-called "sublethal radiation damage" which does not directly result in the irradiated cell's death, but makes it sensitized to the further action of ionizing radiation [35-36].

In a number of cases, instead of an S-shaped survival-dose curve one may observe exponential curve, i.e. the dependence of the form

$$S = \exp(-D/D_0) \qquad (5)$$

Such dependence is a special case of the described model at $n=1$, and D_q in this case is accordingly equal to zero. Such a scheme is referred to as a single-hit single-target model. In this case only one parameter - D_0 - is needed to characterize cells radiosensitivity. Same cells may produce different dose-effect curves (sigmoidal or exponential) depending on the different linear energy transfer (LET) of irradiation. The first type of curves is usually observed upon the action of low-LET radiation, while the second type is observed for high-LET radiation. An important point is that the initial slope of S-shaped dose-effect curves is equal to zero, i.e. the tangent to the curve at the zero dose of radiation is parallel to the X-axis. At the same time, in the radiobiological practice S-shaped curves with non-zero (usually negative) initial slope are frequently observed, with the rest of the curve described by the equation (1). In this case experimental points are best approximated by the so-called modified single-hit multitarget model with the equation of a form

$$S = \exp(-D/_1D_0)\{1 - \exp(-D/_2D_0)]^n\} \qquad (6)$$

It is commonly assumed for such a case that there are two types of cell inactivation in a homogeneous population that can be regarded both to the cells characteristics and irradiation characteristics, or to both factors simultaneously. The mean cell lethal dose D_0 for the exponential region in this case can be derived using the following simple relationship:

$$1/D_0 = 1/_1D_0 + 1/_2D_0 \qquad (7)$$

Another approach to the description of the dose-effect dependence which has gained acceptance in the last decades is based on a so-called linear-quadratic model [36-38]. In this case the dependence of the cell survival on the radiation dose is given by the following relationship:

$$S = \exp[-(aD + bD^2)] \qquad (8)$$

where parameters a and b characterize contributions of the linear and quadratic components into the gradient of the fraction of survived cells (note that many authors use greek characters α and β instead of coefficients a and b; note also the dimensionality of these values – it is Gy^{-1} for the linear coefficient and Gy^{-2} for the quadratic one). The initial slope of such a curve is determined by the a (or α) value, while its following shape at high radiation doses is given by the b (or β) value. It is noteworthy that in this case in the range of high doses the curve "dose-effect" does not tend to the exponential function but is similar to a parabola, i.e. its curvature grows with the radiation dose. When the range of the radiation dose is limited and experimental points are scattered, separate parts of such a curve can obviously be approximated by any dependence, including the exponential one, so the choice of a model depends on the researcher's preferences.

In the first works [16-17], dedicated to the survival of murine hematopoietic stem cells after the action of low-LET radiation (X-rays of 280 kV), it was found that the dose-effect curves for HSC in the exogenous colony test (spleen of lethally irradiated mice 10 days after donor bone marrow transplantation, irradiation in vitro and in vivo) are very similar and have the following parameters characterizing their radiosensitivity: 1. D_0 is 1,05 Gy, and n is equal to 2,5 (when irradiating bone marrow in vitro); 2. D_0 is 0,95 Gy, and n is equal to 1,5 (when irradiating bone marrow in vivo). Similar value of D_0, close to 1 Gy, was obtained in the experiments using the method of spleen endocolonies [36, 39]. Thus, murine hematopoietic stem cells did not differ in their radiosensitivity from the majority of other mammalian cells, which were grown in the culture and whose post-radiation survival was evaluated by change in their ability to form clones [35]. This conclusion about similarity of the radiation reaction of this type of stem cells and mammalian cells cultured in vitro were confirmed by other radiobiological data. The same conclusions were made when the methods of culturing progenitor cells for different hematopoietic lines were developed, and the curves of the radiation survival were obtained [36, 40-42]. The progenitor cells under study included early and late progenitor cells of erythroid series (BFU-e and CFU-e), progenitor cells of granulocytes and macrophages (GM-CFC, G-CFC и M-CFU) and others. As a rule, progenitors differentiated into a certain cell line were slightly more resistant to ionizing radiation than pluripotent stem cells, but this difference was not significant. At the same time, more committed stem cells, but still remaining a source of several hematopoietic lines, are characterized by a higher radiosensitivity as compared to polypotent HSC [43]. Under the action of high-LET radiation (mainly neutrons, protons, heavy charged particles etc.) D_0 for HSC and their committed cell progeny drops by several times (proportionally to the magnitude of the relative biological effectiveness of such radiation – RBE), while n – extrapolation number – decreases to 1.0, i.e. the dose-effect curves convert into exponential functions [36, 44-45]. After the action of high-LET radiation, hematopoietic stem cells, as well as other types of mammalian cells, are almost incapable of repairing radiation damage, as opposed to the action of low-LET radiation [35-36, 42, 46-48]. Similar to mammalian cells in vitro, this type of stem cells was able to repair both sublethal cell damage and potentially lethal damage [35, 46-48]. It was shown later that repair of the damage in HSC genetic structures caused by low-LET radiation along with their proliferation at sufficiently long irradiation are the reasons for the dose-rate effect [36, 48]. HSC and their partially differentiated cell progeny have served as first types of adult stem cells to investigate the effect of radiomodification of various biological, physical and chemical agents, which can either attenuate or enhance their damage from irradiation in vivo and in vitro [21, 36, 49-61]. These studies substantially expanded the knowledge about the nature of fetal and adult stem cells and also about the mechanisms promoting their survival or death in the different whole-organism conditions during fetal and postnatal life, as well as during growth in the conditions of primary and stable cell cultures. Some of these data will be considered in more detail later, in the discussion on the mechanisms of stem cells death in an organism and their importance for sustaining cell homeostasis in the systems of an organism's cell renewal.

After the methods of cell culture and survival estimation were developed for pluripotent HSC and HSC-generated progenitor cells for different large animals and human hematopoietic lines, extensive research was conducted to characterize their radiosensitivity [36, 62-69]. It turned out that, with few exceptions, the quantitative parameters of cell radiosensitivity for these cells were of the same order of magnitude as those for laboratory mice. In other words, for such cells the value of D_0 under the action of low-LET radiation is about 1 Gy, and the values of extrapolation number n are in the range of 2-5 [32, 36]. When applying the "linear-quadratic model" to describe the dose-effect curves for survival of human hematopoietic stem cells, the main quantitative parameter of radiosensitivity – "α/β ratio" - was found to be 3-5 Gy, i.e. of the same order as that for mice [36-37, 70]. Therefore, human HSC cultures as well as laboratory animal stem cells have become a popular object not only for solely radiobiological studies, but also for evaluation of chemotherapy agents especially those involved in the cancer therapy [71-74]. Besides, they are used for evaluation of different modifiers which can be utilized to stimulate the reparative processes in the case of disorders of hematopoiesis in humans [75-76]. Along with the modern molecular biology tests, methods based on the consideration of the death level of stem cells with different degree of committing are still applied very frequently in such studies [65, 77].

Another system of adult stem cells, first investigated by radiobiologists, are intestinal stem cells (ISC) [6, 19-20, 32, 36, 78-80]. As it was for HSC, the first elaborated approaches to estimate ISC survival were developed for mice and by their concept they were analogous to the method of "spleen endogenous colonies". The mice underwent general irradiation in the doses leading to a substantial loss of ISC in the lower region of intestinal crypts. Before the "intestinal" form of radiation death began, 3 days after irradiation, fragments of mice small intestine were fixed for the following histological treatment and obtaining transverse cross-sections of small intestine. The cross-sections were examined for presence of regenerating crypts and those crypts where regeneration had not begun. The estimation was based on a reasonable assumption that regenerating crypts must have one or more ISC remaining after irradiation, while non-regenerating crypts must have no viable ISC, and that the probability to find survived ISC in the small intestine crypts was described by Poisson distribution. Thus, using rather simple mathematics, a mean number of survived ISC per one crypt and also for the entire transverse cross-section in a histological slide could be calculated [19-20, 36, 78]. This allowed one to plot the curve of ISC survival vs. irradiation dose in a certain dose range and to estimate a value of "mean cell lethal dose" D_0 for them. D_0 for murine ISC appeared to be approximately the same as that for HSC and was equal or a little higher than 1 Gy under the action of low-LET radiation [19-20]. However, full curves "dose-effect" for these two types of stem cells are significantly different, since ISC have considerably greater "shoulder dose" D_q. For stem cells of the small intestine epithelium D_q is 4-5 Gy according to different estimations, and the value of "extrapolation number" n is of several tens [78-83]. This fact reflected higher ability of ISC to repair sublethal cell radiation damage (Elkind-type repair) during the early postradiation period, as was soon confirmed by the experiments on the effects of fractionated irradiation for ISC [84-85], and also in the low dose rate experiments [36], when reparative processes in the damaged stem cells could be realized fast. In a number of such radiobiological experiments the "linear-quadratic model" was used to describe the "dose-effect" dependence. According to this model, the "α/β ratio" is 13.3 Gy at a high radiation dose rate (1.2 Gy/min) and is equal to 96 Gy at a low radiation dose rate (0.08 Gy/min) [86]. The conclusion about higher ISC ability to repair radiation damage of low-LET radiation is consistent with the data on a higher biological activity of high-LET neutrons of various energies and other high-LET radiation for these cells as compared to HSC [36, 87-90]. In [91], a relatively infrequently used "two-component" model was applied (see Eq. 6 and 7) to describe dose dependence of ISC survival after γ –irradiation of mice. According to this model, the D_0 value for the first component is 1.5 Gy, for the second component it is 4.5 Gy, with the value of "extrapolation number" n equal to 20. These data also showed better expressed reparative processes for ISC after the action of low-LET radiation, additionally confirmed by a study of fractionation effect and by comparison of effects of Co^{60} γ-radiation and fast neutrons.

The research on the ISC radiobiology allowed creation of a new experimental model for some problems which were mentioned already in the works on HSC, but were difficult to analyze using one object only. First of all, it concerns the concept of "stem cells niche" first formulated for HSC [92] which suggested that the adult cell renewal systems contained certain structures capable, by producing cytokines and growth factors, and also by providing necessary cell interactions, of generating conditions for self-sustainable population of stem cells during lifespan. A concept of "stem cells microenvironment" is widely used along with this term nowadays which extends to all known stem cells types [93], although a detailed research on such structures is only beginning and probably will not be finished soon. This concept is utilized also for analysis of different problems of radiation damage and the following HSC and ISC population regeneration [93-95]. It has been found that survival of ISC in the lower part of mice intestinal crypts after massive irradiation increases significantly when this zone is supplied with FGF-2 produced by pericryptal fibroblasts. The latter have a mesenchymal nature and play a role of "niche" elements for ISC [94]. Thus, the "niche" influence is realized not only in the conditions of sustained cell homeostasis in adult cell renewal systems, but also in the conditions of its disruption resulting from radiation inactivation of stem cells.

Using experimental ISC models of transgenic and knockout mice, researchers successfully investigated the effects of a number of genes onto the apoptosis development and onto radiation damage of stem cells [96-98]. A number of genes (*p53, p21, ATM, Ku80, PARP-1, Msh2 etc.*) directly relevant to DNA repair were shown to play an essential role in the regulation of the programs of cell survival. For example, products of p53 gene served as inducers of apoptotic form of cell destruction in intestinal crypts or led to the growth arrest at cell cycle checkpoints [96, 99]. It was observed for control non-irradiated mice that 1-2 ISC underwent spontaneous apoptosis in the small intestine crypts during 1 day, the fact which was probably related to homeostatic regulation for sustaining the number of stem cells in a crypt, with the participation of p53 genes and genes of the bcl-2 family [96, 99]. Deletion of p53 gene had a differing effect onto ISC survival in mice after low-LET irradiation: their survival in small intestine did not change, though regeneration of damaged crypts slowed down, while in large intestine ISC survival was lower for p53(-/-) mice than that for the wild type mice [98]. It was believed that in this case attenuation or absence of stem cells arrest in G_2-phase of cell cycle did not give them enough time for the reparative processes to go and resulted in the following decrease in cell survival. Under the action of high-LET radiation p53-independent apoptosis was induced in ISC [100]. No difference in the extent of apoptosis within the whole pool of intestinal crypt cells was found between adult wild-type (WT) and p21(-/-) mice, but p21(-/-) mice showed 3 times higher crypt survival than WT mice 3.5 days after irradiation in the dose of 13 Gy [101]. The increase in the survival of ISC and their proliferating and differentiating cell progeny in irradiated animals was associated presumably with increased numbers of Msi-1- and survivin-expressing cells in regenerative crypts. Besides, it was shown that p53 gene products participated in the development of early apoptosis induced by very low radiation doses in ISC located in the base of small intestine crypts [102]. Late manifestations of radiation apoptosis for ISC are p53-independent [99, 102]. Although no change in the probability of spontaneous or radiation-induced apoptosis was observed for mice with the deletion of bcl-2 gene family, but *bad* gene products were expressed in the crypt cells and villi after the action of radiation [102]. An unexpected phenomenon of significant increase in radioresistance of ISC was revealed for the cells located in the zone of patches of Peyer upon the action of both low-LET and high-LET radiation [103-104]. It was suggested that the increase in the ISC radioresistance in this case was due to "physiological shut down" of the p53 gene activity and/or increased expression of the bcl-2 and bcl-x genes [105].

A unique property has been discovered for ISC of small intestine – preservation of stability of two template DNA strands after cell division in the remaining stem cell, while the second cell going into differentiation receives two newly synthesized DNA strands [106-107]. Such segregation of template and newly synthesized DNA strands in stem cells of small intestine provides a peculiar "immortality" of DNA in this type of adult stem cells and stability toward development of cancer that makes them strikingly different from stem cells of large intestine epithelium. At the same time, it is worth mentioning that radiobiological aspects of many signaling pathways regulating cell homeostasis, including cell proliferation, cell death, possible carcinogenesis (such as Wnt/β-catenin, BMP/SMAD4, Notch, Hedgehog, PTEN/aKt, TGF- β etc.) have been stud concerns effects of different modifiers of stem cells' radiosensitivity, with initiation of a variety of these signaling pathways. For example, epithelial growth factors (TGF-β, KGF) and cytokines related to hematopoiesis and immune functions (IL-1, SCF и IL-11, IL-12), usually in the form of human recombinant polypeptides, appeared to be powerful exogenous modifiers of ISC survival/death [108]. Interleukin α (IL-1α) was one of the first cytokines which was found to have radiomodifying properties toward ISC [109]. Administration of IL-1α to animals 4-8 hrs before irradiation resulted in the aggravation of crypts damage. Only if administered 1-7 days before irradiation, it led to a decrease in the ISC death, though repopulation of the damaged crypts was not faster [110]. It has been known that IL-1α radioprotective action onto HSC also appears in case of preliminary injection performed several hours or days before irradiation [111]. Interleukin-12 (IL-12), secreted under the action of bacterial products by monocytes/macrophages and lymphocytes plays an important role in the generation of T1-helper cells. With other growth factors, it synergetically stimulates proliferation of early hematopoietic progenitor cells thus providing radioprotective action [112]. At the same time, this cytokine is capable of suppressing tumor growth and metastases production, probably by inhibiting cancer stem cells, and also of decreasing the proliferative activity of ISC [112-113]. This fact is demonstrated by a significant drop in the number of regenerating crypts and a twofold decrease in the lifespan of mice having received this cytokine in the dose of 15 Gy either 18 hrs before irradiation, or 1 hr after irradiation. Antibodies to interferone-gamma (aIFN-γ) cancelled this sensitizing effect [113]. Since a controversial action of IL-12 onto the radiation effect for different tissues was found, the prospective of its application as well as application of similar cytokines as radioprotectors seems to be somewhat illusive. Radioprotective action of FGF-2 was noticed earlier via the test of ISC survival [94]. An extensive study on the mechanisms of this effect has shown that they are not associated with direct action of FGF-2 onto the cells of intestinal epithelium, since its receptors are present only in the endothelium of microvessels surrounding crypts [114]. In the endothelial cells FGF-2 inhibited activity of acid sphingomyelinase producing a well-known thanatogenic mediator – ceramide. FGF-2 receptors are not expressed in the intestine, while acid sphingomyelinase is presented in the vessels endothelium in the amount which is 20 times higher than that in other tissues. Taking the above facts into account, it was suggested that the primary cause of intestinal mucosa exhausting and, correspondingly, ISC death, was early apoptotic destruction of the microvessels endothelium. Morphologically, FGF-2 radioprotective action was expressed in the limiting of crypts shrinking, but not in the acceleration of their regeneration after irradiation [114]. At the same time, it should be mentioned that the extent of the radioprotective effect for FGF-1 and FGF-2 depends greatly on the experimental mice strain [115]. Transforming growth factors (TGFβ1, TGFβ2, TGFβ3) are known as inhibitors of epithelial cells proliferation, arresting them in the G_1-phase of the cell cycle, thus they are capable of affecting ISC radiosensitivity [116]. For example, exogenous TGFβ3, injected to mice 24, 8, 4 hrs or immediately before irradiation, essentially increased the number of regenerating crypts in the small intestine (3-12 times increase, depending on the radiation dose). In the large intestine, the effect was less noticeable, and the number of regenerating crypts increased no more than 2.5 times [116]. A similar radioprotective action was found for the keratinocytes growth factor (KGF) which is a member of the fibroblast growth factor family (FGF-7). KGF stimulates reparative processes, as was shown in the models of skin wounds and experimentally induced colitis [117-118]. KGF radioprotective effect onto intestinal epithelium is observed only if it is administered before irradiation; the number of regenerating crypts increases by a factor of 3.5 for the 14 Gy radiation dose. Since its receptor is expressed in any epithelial cells, KGF presumably directly affects ISC survival, possibly via the induction of seleno-independent glutathione-peroxidase activity [118]. However, it should be pointed out that intestinal crypts of glutathione-peroxidase deficient mice ($Gpx1^{-/-}$) were more resistant to radiation, than those of the wild-type mice [119]. KGF radioprotective action may also be related to the increase of ISC population and/or their accumulation in the radioresistant S-phase of the cell cycle [102, 120]. A combination of KGF and stem cell factor (SCF) did not additionally boost ISC radioresistance [121], though individually applied SCF improved survival of stem cells of this type in irradiated mice [122]. This fact may indirectly show that the radioprotective effect of these agents is realized via the same mechanisms.

Another class of extensively studied anti-radiation agents for adult stem cells is lipopolysaccharides (LPS). LPS, components of a cell membrane of gram-negative bacteria, are one of the most well-known and powerful modifiers of various mammalian cells (including HSC and ISC) survival upon irradiation in vivo [52, 123]. Note that the important details of the radioprotective action of LPS onto ISC have been found only in the last decade [124]. The regulatory circuit appears as follows: introduction of LPS increases expression of the factor of tumor necrosis alpha (TNF-α) in the small intestine (exact cell origin is not identified) by almost 4 times; TNF-α binds to TNFR1 in pericryptal fibroblasts and/or villus enterocytes and promotes synthesis of cyclooxygenase Cox-2 and prostaglandin PGE_2 in these cells. RNA-binding protein Apobec-1 is found in the same chain [125]. It should be noted that PGE_2 is not synthesized under the action of LPS in the cells forming a crypt and carrying TNFR1. Such an effect is observed also for murine enterocytes expressing mutant Apobec-1. The increase in ISC survival due to prostaglandin may be relevant to the inhibition of the mechanisms of apoptotic destruction and to the arrest of irradiated stem cells in the G_2-phase that gives them time for the reparative processes [126]. It is not improbable that LPS ability to inhibit apoptosis via suppression of p53 activity plays a certain role in its antithanatogenic action [127]. In general, it is conventional to consider the radioprotective effect of LPS onto stem cells and their microenvironment as a result of the LPS-induced prostaglandins production [124, 128].

Among the low-molecular stimulators of ISC post-radiation survival, dimethylsulfoxide (OH$^{\bullet}$-radicals scavenger) and retinoic acid (a well-known inducer of cell differentiation) were found [129]. cAMP-phosphodiesterase inhibitors (diethylamino-1-reserpine, 1-methyl-3-isobuthyl-xanthine, theophylline and caffeine) administered to mice shortly before irradiation boosted the number of regenerating crypts by 6-7 times in respect to the control group [130]. The effect in this case did not visibly differ from that for the standard radioprotector WR-2721, which acts at the physico-chemical level via decreasing the degree of DNA radiation damage in various types of adult stem cells [131-132]. Two more groups of substances, opposite in their general effect, are of special interest for radiomodification of stem cells. These two groups include anti-inflammatory agents (by the example of indometacin) promoting death of irradiated ISC, and carcinogens (by the example of azomethane and 1,2-dimethylhydrazine) promoting survival of irradiated stem cells [126,133-135]. For example, indometacin, non-selective inhibitor of cyclooxygenases 1 and 2 (Cox-1, Cox-2), decreases the concentration of prostaglandin PGE_2 both in intact and in irradiated animals, though the latter ones show a rise in the Cox-1 level after irradiation. Administration of this drug to the animals 1 hr after irradiation and then every 8 hrs for 3 days resulted in the dramatic reduction in the number of regenerating crypts [133]. Selective Cox-2 inhibitors did not affect the crypts' survival, as well as the deficit of these genes (Cox-$2^{(-/-)}$) in irradiated mice. However, neutralizing antibodies to PGE_2 and Cox-$1^{(-/-)}$-genotype decreased the number of regenerating crypts [136]. As the analysis of these data has shown, prostaglandins switch into the mechanisms of ISC death only in the damaged epithelium, in the conditions of stress reaction onto cytotoxic action [133]. In contrast, carcinogens - azomethane and 1,2-dimethylhydrazine – administered to mice 1 day before irradiation were able to essentially enhance ISC survival as compared to the control group, where the animals did not receive carcinogens [134-135]. This effect may be related to the damage brought by the carcinogens to the mechanisms of apoptotic removal of stem cells containing radiation-damaged DNA, for example, via inhibiting the activity of p53 gene. At the same time, it was shown using the ISC model that combined application of a carcinogen and indometacin in irradiated mice attenuates radioprotective action in comparison with the effect of the carcinogen only [134], i.e. prostaglandins participate in the realization of carcinogens radioprotective action, either. We have shown recently that 1,2-dimethylhydrazine is capable of exhibiting the same radioprotective action onto murine HSC [135], irradiated in vivo, though additional indometacin administration did not suppress this effect. It is possible that a specially synthesized p53 gene inhibitor - pifithrin-α - has a radioprotective action onto adult stem cells similar to the carcinogens effect [137]. On the other hand, in the mechanisms of pifithrin-α and its analogues action, their ability to inhibit NO-synthase activity may play a certain role, which is characteristic for many radioprotectors with physico-chemical mechanism of action [55].

Rafiomodifying action of the majority of agents investigated mainly using HSC and ISC remains absolutely insufficiently studied for other types of adult stem cells. Among those, we earlier mentioned mesenchymal stem cells (MSC) [6, 23-28], stem cells of spermatogonial epithelium and other epithelia [29-31, 34], and also so-called cancer stem cells [138-140]. At the same time, even those few radiobiological studies conducted with the use of the above types of stem cells allowed finding a number of interesting phenomena. For example, it was revealed that MSC of bone marrow were characterized by a higher radioresistance than HSC of bone marrow in vitro and in vivo [24, 27-28, 36, 141-143]. The magnitude of the mean cell lethal dose D_0 for low-LET irradiation of human and laboratory animals' MSC is 1.4-2.0 Gy and can be even higher (up to 2.5 Gy) after prolonged irradiation [144]. MSC population obtained by cell culture of rats' bone marrow is inhomogeneous and contains fibroblastoid precursor cells forming "compact" and "diffuse" clones which significantly differ in their radiosensitivity [143]. If the linear-quadratic model for the description of the survival-dose dependencies is used for these two MSC subpopulations in vitro, the "α/β ratio" is 12.7 ± 5.5 Gy for precursor cells forming "diffuse" colonies, and is 4.5 ± 3.0 Gy for precursor cells forming "compact" colonies. These data can be explained on the hypothesis that precursor cells forming "compact" colonies belong to a category of more "primitive" and less committed MSC, while more radioresistant precursor cells forming "diffuse" colonies belong to more committed cell progeny, with partially exhausted proliferative potential. This assumption is consistent with the data obtained in our laboratory on the MSC cultures of patients with Hodgkin's lymphoma. In these patients, large portions of bone marrow underwent intense radiotherapy resulting in a high proliferative load onto different populations of bone marrow cells. We have shown that the MSC cultures grown from non-irradiated regions of the patients' bone marrow contain a greater fraction of "diffuse" colonies [145]. We have also found in the experiments on animals that the "oxygen effect" is expressed much weaker for irradiation of this MSC subpopulation in vivo, i.e. these cells are under hypoxia conditions in an organism [146]. There is no question that MSC and ISC remain a good object for studying cell nature of radiation aging [147-148].

As we have recently discovered in the experiments on 5-azacytidine-induced human and rats' MSC differentiation toward cardiomyocytes, their radioresistance increased [149]. This fact is in a good agreement with the general understanding that this is a characteristic phenomenon in the process of pluripotent adult stem cells differentiation into progenitor cells of certain cell lineage. It would be interesting to observe how MSC radioresistance changes when they differentiate into other numerous cell lineages (for example, progenitor cells for osteoblasts, chondrocytes, adipocytes etc.), but such data have been non-existent so far. Another prospective direction in the radiobiological studies with the use of MSC may be investigation of radiosensitivity of gene-modified MSC cultures, as it is this cell type that is believed to be the most promising agent for gene delivery into different tissues [150-151]. Radiobiological data could be useful in the development of optimal schemes for cell and gene therapy applying MSC and their partially differentiated cell progeny, while the data quantitatively characterizing these cells may be utilized for a peculiar "quality control" of cell cultures industrially manufactured for therapeutic purposes.

A very interesting phenomenon was detected in the radiobiological studies using spermatogonial stem cells - cells existence in the conditions of physiological hypoxia in an intact organism, the fact making this type of adult stem cells relatively radioresistant (the value of D_0 for low-LET radiation is about 1.6-2.4 Gy [29, 36, 152]). Spermatogonial stem cells showed fast increase in the sensitivity to repeat irradiation due to reoxygenation in the first hours after the first action of ionizing radiation, and also efficient radiosensitization in the presence of electron affinic substances [152-154]. This pool of adult stem cells is characterized by exponential decrease in its size with age, in contrast with stem cells in other systems of cell renewal [155]. A local hyperthermic action (in the range of $41-43^0$ C, 30 min long) onto mice testicles before irradiation significantly aggravated radiation damage of spermatogonial stem cells [156]. A research has been started on the "niche" for this interesting system of adult stem cells [157], which provides both self-maintenance of the stem cell pool and their following differentiation up to the spermatozoons production. Although the key role of the Sertoli cells in this system functioning is of no doubt, studies on the specific programs of interaction of different components of this cell renewal system and their reaction onto damaging agents are in their beginning stage only.

Very little radiobiological information exists on the characteristics of cancer stem cells. These cells are traditionally considered as relatively resistant to ionizing radiation and many chemotherapy drugs [158-160], but the nature of this resistance remains almost unknown. This might be related to the fact that they exist in the conditions of hypoxia in an organism, or they can have p53 function shut down or genes of bcl family activated, they can possess a powerful system of reparative enzymes etc. Study of these mechanisms for developing radio- and combined therapy for patients with resistant forms of malignancies has become an urgent task. It is not impossible that the model based on the observation of radiobiological characteristics of adult stem cells for animals treated with carcinogens will become one of the most convenient models, as mentioned earlier [134-135]. Both expanding the set of adult stem cells types (first, by using MSC and their partially differentiated progeny) and monitoring the effects not only immediately after the carcinogen administration but also in a more distant time appear to be promising.

Stem cells within the "critical" systems of cell renewal which preserved their viability after total irradiation are currently considered as certain "determinants" of an organism's survival in the acute phase after a lethal dose irradiation leading to "bone marrow" and "intestinal" forms of radiation lethality [36, 161]. This is due to the fact that they are the only sources of proliferative restoration of the corresponding systems of cell renewal devastated by radiation. In this case the adult stem cells which survived radiation escaped the death which manifests itself in the form of radiation-induced apoptosis or mitotic catastrophe. Radiobiological analysis performed using "intestinal" form of mice death has shown that at such radiation doses an average of one or less stem cells survives in the crypts of small intestine. These surviving stem cells may restore the initial cell population of a crypt relatively fast via proliferation [16]. Those intestinal crypts where no stem cells survived after irradiation still may restore their structure due to the effect of "fission" of nearby repopulated crypts [162]. For the hematopoietic system the restoration of hematopoiesis is possible due to the phenomenon of HSC "migration" [163-164]. Death and damage of stem cells also play a crucial role in the development of late non-tumorous radiation damage of different tissues [4]. However, other effects, such as still poorly investigated effects of radiation-induced senescence of stem cells, as well as accumulation of somatic mutations and clonal stabilization, may contribute into the pathogenesis of late radiation damages [147-148]. As for the development of radiation-induced tumors, epigenic disorders and the possibility of the "plasticity"effect (dedifferentiation) may greatly contribute along with somatic mutations in stem cells and disorders in their "microenvironment" [165-166], but the real progress in this direction is expected only in the future research.

Chapter 3

HYPERTHERMIA AND ADULT STEM CELLS

Going to the description of the effect of another physical factor - hyperthermia – capable of lethally damaging adult stem cells in vitro and in vivo, we emphasize that thermal biology of stem cells has been poorly developed so far. The research in this direction was stimulated by first attempts of utilizing methods of general and local hyperthermia to boost the efficiency of radiotherapy for resistant malignancies in the second half of XX century [167-168]. As has been established in the experiments on the biological reasons for such hyperthermia application, heating of normal and tumor cells of humans and laboratory animals in vitro and in vivo to 40-41^0 C and more increases their sensitivity to ionizing radiation, besides, heating has its own damaging action onto cells [169-170]. The damaging and radiosensitizing action of hyperthermia at the cell level depends on the temperature and duration of heating, and also on such factors as cells' nature, pH of the medium etc. Hyperthermia is capable of aggravating damaging action of not only ionizing radiation, but also of many chemotherapy drugs used in the cancer treatment [171]. One rather unusual phenomenon is characteristic for the biological action of hyperthermia as compared to the majority of other damaging agents – development of a higher resistance to the repeat hyperthermia observed in a short (1-2 days) time after a mild heating. This phenomenon was called "induced thermotolerance" [169-172]. The basis for the thermotolerance phenomenon is induction of heat shock proteins and a raise in the activity of DNA polymerase-beta (a key enzyme; suppression of its activity under hyperthermia leads to cell death) [173]. Heat shock proteins are known to perform important functions in the body on sustaining homeostasis and providing reparative processes. They are synthesized not only in response to hyperthermia, but also under the action of many stress agents [174-176]. Therefore, it seemed interesting to investigate damaging and radiosensitizing action of hyperthermia and its ability to induce thermotolerance in adult stem cells. It was shown in the experiments conducted on murine HSC heated in vitro that hyperthermic "dose-effect" curves plotted in a manner similar to that for radiation (X-axis is the duration of heating for each of 4 temperatures in the range of 41-44^0 C, Y-axis is the logarithm of survived fraction of cells) had a typical sigmoidal shape with a small "shoulder dose" and the subsequent exponential region [177]. This finding means that the "single hit multi-target" model suggested in radiobiology can be used for the description of the dependence of adult stem cells survival on the "heat dose". The latter is estimated by the duration of heating at a given temperature. The D_0 for murine HSC heated in vitro and then transplanted to lethally irradiated mice for the survival estimation using spleen "exocolonies" was found to be 29.3, 22.6, 8.1 and 2.8 min for the temperatures of 41^0, 42^0, 43^0 and 44^0 C, respectively. An additional study on the distribution of spleen colonies by their morphological forms showed that it did not differ statistically significantly for stem cells survived hyperthermia and for control intact HSC. In other words, the choice of differentiation for stem cells after hyperthermia in the studied range of temperatures and durations of heating was not disrupted. The analysis of the character of D_0 change with temperature showed that this dependence was consistent with the second order kinetics of chemical reactions describing reaction rate vs. temperature [169-170, 178]. The main parameter of such dependence – "activation energy" - was estimated to be about 120-150 kcal/mol for the process of heat inactivation of murine HSC. This value is close to similar parameters for other mammalian cells (but not stem cells) in vitro and in vivo [178-179], i.e. similar mechanisms can be assumed for the damaging action of hyperthermia onto adult cells with different functions. In the same work the radiosensitizing action of hyperthermia onto murine HSC was described. The cell suspension was heated to 43^0 C for 30 min that decreased stem cells survival up to 10% of the initial value. The subsequent irradiation of the preliminary heated suspension resulted in the reduction of HSC radiosensitivity according to the test of clonogenic survival with the D_0 being decreased from 0.97 Gy to 0.61 Gy, or about 1.5 times. Approximately the same degree of HSC radiosensitization was observed if the cells were heated 1 hr before irradiation. However, when the HSC were heated 1 hr after irradiation, the observed reduction in the cell survival was only a simple summative effect of radiation and heating. These data can be explained as a result of fast reparative processes in the irradiated stem cells which can be blocked only by preliminary heating. The latter conclusion is confirmed by the data on the hyperthermia blocking the activity of DNA repair enzymes, in particular, DNA-polymerase-beta [173, 178, 180]. Similar results on the thermal inactivation and radiosensitization were obtained by our group and other authors for other types of adult stem cells of humans and laboratory animals – progenitor cells of granulocytes-macrophages [181-182], spermatogonial stem cells [156, 183], mesenchymal stem cells [175, 182, 184]. The general conclusion based on these data is that the prepared suspensions of adult stem cells are characterized by almost the same sensitivity toward hyperthermia, while transplanted tumor or leukemic cells often demonstrate higher sensitivity to the heating [168, 179, 182, 185-186]. For the murine leukemic myeloid cell line L1210, it was possible to bind their elevated heat sensitivity to a disorder in the balance of protein expression for the Bcl-2 family, with inclination toward the family members possessing proapoptotic action [186]. At the same time, there are indications that in a tumor tissue some cells can be different in their heat sensitivity which was demonstrated in the study of this sensitivity for several different cell clones obtained from human colon adenocarcinoma [187]. It remains unclear how precisely the data of testing adult stem cells thermosensitivity in vitro correspond to their thermosensitivity in vivo. In our works performed on MSC of rat bone marrow, we obtained "dose-effect" curves for the 43^0 C heating of a prepared suspension of bone marrow cells and for a local heating of a lower extremity (by a controlled microwave radiation). We found that the MSC thermosensitivity in vitro and in vivo did not noticeably differ [182, 184]. In these experiments we also discovered a possibility to induce MSC thermotolerance in vivo via local action of microwave radiation. This allowed us to create a method of temporary "labeling" of stem cells in an organism to observe certain physiological processes, for example, stem cells migration into other parts of the body upon their damage or when using different actuators of cell migration. These data appear to be of top interest in connection with the development of cell therapy often supplemented with stimulation of autologous bone marrow stem cells migration into various damaged organs and tissues [188].

Taking into account an important role of p53 gene in apoptosis, attempts were made to ascertain this system participation in the hyperthermic death and radiosensitization of normal and tumor cells in the whole organism conditions. It was shown that hyperthermic sensitization for different types of cells which underwent irradiation with different LET values manifested itself in the form of activation of apoptotic processes and possibly necrosis regulated by expression of p53, Bcl-2 and Bax [189-193]. The damaging action of hyperthermia alone was found to have no close relation to the activity of p53 gene and was observed for the mutant cells or in the absence of this gene [194-195]. This may be regarded to the fact that a medium-efficiency heat dose may disrupt the process of apoptosis, and also in this case necrosis processes may be realized [196]. Thus, the analysis of the realization of the hyperthermia mechanisms at the cell level and its application in the new methods of anticancer and cell therapy remains an urgent task of the modern biomedical research.

Chapter 4

PHENOMENON OF "ISCHEMIA/REPERFUSION" FOR ADULT STEM CELLS

One more interesting problem in the field of agents acting on adult stem cells in vivo is an "ischemia/reperfusion" reaction recently discovered by our group for stem cells of two "critical" systems of a whole ogranism's cell renewal [197]. This reaction is known to result in cell apoptosis or necrosis in different tissues of a whole organism after acute hypoxia caused usually by temporary disruption in the delivery of oxygen and other necessary components for energetic and plastic demands of tissues. Until recently, it has been investigated only for highly differentiated, usually non-proliferating parenchymal cells of vitally important organs (heart, kidneys, nervous system etc.) [198-200]. A possible development of this reaction in the stem cells within the tissues designed by the principle of cell renewal almost did not attract attention of researchers. However, stem cells in such tissues (primarily, in bone marrow and small intestine epithelium) play a crucial role both in sustaining their physiological regeneration and in the response to the action of damaging agents, especially ionizing radiation and cytostatics [36]. Besides, it has become clear in the last years that the regeneration processes in an organism with damaged vitally important organs (including the case of "ischemia/reperfusion") can be markedly enhanced by delivery of MSC into the "target" tissues. This can be realized through transplantation of autologous bone marrow MSC or by activation of own MSC migration [201-203]. Therefore, we projected a search for indications of the reality of the "ischemia/reperfusion" reaction in stem cells of two "critical" systems of cell renewal (bone marrow and intestinal epithelium). The cells survival was studied by radiobiological methods after total irradiation using a known radioprotector serotonin to generate a short-lasting acute hypoxia in the tissues during irradiation [204-205]. Sodium nitroprusside (SNP) was used as an agent for testing the "ischemia/reperfusion" reaction in stem cells. SNP is a donor of NO radicals and is known to significantly enhance the cells' probability to survive when administered to animals in the period of "reperfusion (reoxygenation)". NO radicals compete with active forms of oxygen (AFO) and thus diminish their damaging action [198, 206]. In the experiments on the mice from the control group which underwent only total irradiation in the dose of 6 Gy, the number of endogenous spleen colonies was 1.9±0.2 colonies/spleen on the 8th day after irradiation. The administration of hypoxic radioprotector serotonin to mice 10 min before irradiation increased this level to 6.1±0.5 colonies/spleen, i.e. by about 3 times. This was consistent with the results of similar experiments on the evaluation of "bone marrow" survival of mice or on the HSC survival [36, 204, 207]. However, when SNP was administered to serotonin-protected mice immediately after irradiation, it resulted in the further essential growth of the number of endogenous spleen colonies to 13.2±0.7 colonies/spleen on the 8th day after irradiation. At the same time, administration of SNP alone to mice did not produce radioprotective effect, and the number of registered endogenous spleen colonies was only 2.2±0.3 colonies/spleen which did not differ statistically significantly from the control group. The discovered effect of the strong radiomodifying action of the post-radiation SNP administration to mice after prior protection with serotonin before irradiation was quite unexpected and was obtained for the first time for adult stem cells. This effect can be explained as an additional suppression of the "ischemia/reperfusion" reaction in HSC irradiated in the conditions of rather substantial but short-term hypoxia. It may be suggested that the real anti-radiation protection due to acute hypoxia in vivo is actually higher than the observed one, but the majority of survived stem cells (about 50%) undergo apoptosis or necrosis at the stage of reoxygenation because of the "ischemia/reperfusion" reaction. The validity of this hypothesis was tested in the experiments where the post-radiation survival of HSC with different degree of committing (CFU-S-8 and CFU-S-12) was determined by the method of exogenous spleen colonies. In this case the same groups of mice as those in the experiment with exogenous spleen colonies were used as donors of bone marrow. However, the total mice irradiation was conducted in the dose of 2 Gy, and the bone marrow of these animals was obtained 2 hrs after irradiation, when the completion of the "ischemia/reperfusion" reaction in HSC was expected. The number of colonies (formed by the remaining live HSC) was measured grown in the spleen 8 and 12 days after transplantation of a given amount of donor bone marrow cells into lethally irradiated recipients. The level of the counted spleen colonies was normalized to the number of transplanted cells of bone marrow. The stem cells survival was calculated in comparison with the survival of the same cells from non-irradiated animals. The "ischemia/reperfusion" reaction was also established in this case, being more expressed for CFU-S-8 than for CFU-S-12. This may be related to the fact that more "primitive" CFU-S-12 are likely to exist in an organism in the conditions of relative physiological hypoxia and proliferative resting thus being less vulnerable to the "ischemia/reperfusion" reaction. These considerations are consistent with the previously reported data on the radiobiology of these two HSC subpopulations [208-209]. A similar picture of the enhancement of the serotonin radioprotective effect by the post-radiation SNP administration was also detected in the test of the intestinal stem cells survival using the method of intestinal "microcolonies". Note that the radioprotective action of serotonin for the stem cells of small intestine epithelium was less marked, and the increase in the survival due to additional SNP after initial serotonin protection was lower, correspondingly. It is possibly related to a lower degree of hypoxia created by serotonin in the intestine, and also to a shorter duration of hypoxia than that in the hematopoietic tissues [210]. The same enhanced radioprotective effect at additional SNP administration was detected in the test on the mice survival for the "bone marrow" and "intestinal" form of death, which reflects the importance of the "ischemia/reperfusion" phenomenon in the organism reaction to the acute radiation damage.

Thus, using radiobiological methods for estimation of adult stem cells survival, for the first time it has become possible to reveal the "ischemia/reperfusion" reaction for such cells in vivo. The primary tasks for the further research in this direction involve elucidating the dynamics of this reaction in the whole organism and study of different modifiers for this reaction. In the latter task, the presence of many inducers of apoptosis and necrosis should be considered, which were found in the experiments on the cell death in tissue culture or in the short-lasting cultures prepared from specially separated parenchymal cells of vitally important organs [211]. From this viewpoint, a "preconditioning" technique is of importance when using adult stem cells (primarily, MSC) for their systemic transplantation into a damaged organism. The "preconditioning" is essential so that transplanted cells would be able to avoid attack of active forms of oxygen after they reach the damaged parts of tissues in the process of "homing" [212-214]. Our scheme of experiments on the studying the "ischemia/reperfusion" reaction may become useful for studying various ways of "preconditioning" and other methods for creation of favorable conditions to sustain stem cells survival after their transplantation into an organism. On the other hand, it seems important to attempt applying the "ischemia/reperfusion" reaction to destroy cancer cells or at least increase their sensitivity to radiation of chemotherapy drugs. This experiment can be performed using the previously considered model of adult stem cells treatment with carcinogens in vivo.

ACKNOWLEDGMENTS

We thank Svetlana Kotova for the help during work with this book.

REFERENCES

[1] Repin V.S., Sukhikh G.T. Medical cell biology. *BEBM publ.*, M., 1998, 200 pp. (Russian).
[2] Korbling M., Estrov Z. Adult stem cells for tissue repair - a new therapeutic concept? *N. Engl. J. Med.* 2003; 349(6): 570-82.
[3] Michalowsky A. The pathogenesis of the late side-effects of radiotherapy. *Clinical Radiology.* 1986; 37(3): 203-7.
[4] Konoplyannikov A.G. Molecular and cellular mechanisms of late radiation injuries. *Radiats. Biol. Radioecol.* 1997; 37(4): 621-8. (Russian).
[5] Turksen K. (Ed.) Adult stem cells. Humana Press Inc., Totowa, New Jersey, 2004, 346 pp.
[6] Potten C.S., Clarke R.B., Wilson J., Renehan A.G. Tissue stem cells. Taylor and Francis Group, 2006, 404 pp.
[7] Blau H.M., Brazelton T.R., Weimann J.M. The evolving concept of a stem cell: entity or function? *Cell.* 2001; 105(7): 829-41.
[8] MacArthur B.D., Please C.P, Oreffo R.O. Stochasticity and the molecular mechanisms of induced pluripotency. *PLoS ONE.* 2008; 3(8): e3086.
[9] Anversa P., Leri A., Kajstura J. Cardiac regeneration. *J. Am. Coll. Cardiol.* 2006; 47(9):1769-76.
[10] Rao M.S. (Ed.). Neural development and stem cells. Humana Press Inc., Totowa, New Jersey, 2006, 454 pp.
[11] Penn M.S.(Ed.). Stem cells and myocardial regeneration. Humana Press Inc., Totowa, New Jersey, 2007, 316 pp.
[12] Coulombel L. Identification of hematopoietic stem/progenitor cells: strength and drawbacks of functional assays. *Oncogene.* 2004; 23(43): 7210-22.

[13] Wright N.A. Epithelial stem cell repertoire in the gut: clues to the origin of the cell lineages, proliferative units and cancer. *Int. J. Exp. Patol.* 2000; 81(2): 117-43.
[14] Dürr M, Müller AM. Plasticity of somatic stem cells: dream or reality? *Med. Klin. (Munich).* 2003; 98(Suppl 2): 3-6.
[15] Shojaei F., Menendez P. Molecular profiling of candidate human hematopoietic stem cells derived from human embryonic stem cells. *Exp. Hematol.* 2008; 36(11): 1436-48.
[16] Till J.E., McCulloch E.A. A direct measurement of the radiation sensitivity of normal mouse bone marrow cells. *Radiat. Res.* 1961; 14(2): 213-22.
[17] McCulloch E.A., Till J.E. The sensitivity of cells from normal mouse bone marrow to gamma radiation in vitro and in vivo. *Radiat. Res.* 1962; 16(6): 822-32.
[18] Till J.E., McCulloch E.A. Repair processes in irradiated mouse hematopoietic tissue. *Ann. N.Y. Acad. Sci.* 1964; 114(1): 115-25.
[19] Withers H.R., Elkind M.M. Dose-survival characteristics of epithelial cells of mouse intestinal mucosa. *Radiology.* 1968; 91(5): 998-1000.
[20] Withers H.R., Elkind M.M. Microcolony survival assay for cells of mouse intestinal mucosa exposed to radiation. *Intern. J. Radiat. Biol.* 1970; 17(3): 261-7.
[21] Metcalf D., Moore M.A. Haemopoietic cells. London, Acad. Press, 1971. 540 pp.
[22] Ploemacher R.E. Characterisation and biology of normal human haematopoietic stem cells. *Haematologica,* 1999; 84: 4-7.
[23] Friedenstein A.J., Chailakhjan R.K., Lalykina K.S. The development of fibroblast colonies in monolayer cultures of guinea-pig bone marrow and spleen cells. *Cell Tissue Kinet.* 1970; 3: 393-403.
[24] Friedenstein A.J., Lurya E.A. Cell basis of haematopoietic microenvironment. M., Medicina, 1980, 216 pp. (Russian).
[25] Caplan A.I. Mesenchymal stem cells. *J. Orthop. Res.* 1991; 9: 641-50.
[26] Pittenger M.F., Mackay A.M., Beck S.C., Jaiswal R.K., Douglas R., Mosca J.D., Moorman M.A., Simonetti D.W., Craig S., Marshak D.R.. Multi-lineage potential of adult human mesenchymal stem cells. *Science.* 1999; 284(5411): 143-7.
[27] Konoplyannikov A.G., Rudakova S.F. Radiosensitivity of guinea pig bone marrow cells forming fibroblast colonies in monolayer cultures. *Radiobiologiia.* 1973; 13(1): 138-40. (Russian).

[28] Latsinik N.V., Sidorovich S.Iu., Gorskaia Iu.F., Pronin A.V., Keĭlis-Borok I.V. Radiosensitivity and the postradiation changes in bone marrow stromal colony-forming cells. *Radiobiologiia.* 1979; 19(6): 848-57.(Russian).
[29] Withers H.R., Hunter N., Barkley H.T. Jr, Reid B.O. Radiation survival and regeneration characteristics of spermatogenic stem cells of mouse testis. *Radiat. Res.* 1974 ; 57(1): 88-103.
[30] Konoplyannikova O.A., Konoplyannikov A.G. Radiosensitivity of stem cells in the spermatogenic epithelium of mice of different strains and different ages. *Radiobiologiia.* 1988; 28(1):31-5. (Russian).
[31] Malcahy R.T., Gould M. N., Clifton K.H. The survival of thyroid cells: in vivo irradiation and in situ repair. *Radiat. Res.* 1980; 84(3): 523-8.
[32] Poten C.S., Hendry J.H. (Eds.) Cytotoxic insult to tissue. Effects on cell lineages. Churchill Livingstone, Edinburgh-L.-Melbourne-NY, 1983, 422 pp.
[33] Coderre J.A., Morris G.M., Micca P.L., Hopewell J.W., Verhagen I., Kleiboer B.J., van der Kogel A.J. Late effects of radiation on the central nervous system: role of vascular endothelial damage and glial stem cell survival. *Radiat. Res.* 2006; 166(3): 495-503.
[34] Jirtle R.L., Michalopoulos G., McLain J.R., Crowley J. Transplantation system for determining the clonogenic survival of parenchymal hepatocytes exposed to ionizing radiation. *Cancer Res.* 1981; 41(9 Pt 1): 3512-8.
[35] Elkind M.M., Whitmore G.F. The radiobiology of cultured mammalian cells. Gordon and Breach Sci. Publ., N.Y.-London-Paris, 1967, 616 pp.
[36] Konoplyannikov A.G. Radiobiology of stem cells. M., Energoatomizdat, 1984, 120 pp. (Russian).
[37] Bloomer W.D., Adelstein S.J. The mammalian radiation survival curve. *J. Nucl. Med.*, 1982; 23(3): 259-65.
[38] Zimmer K.G. Studies in quantitative radiation biology. Oliver and Boyd, L., 1961. 88 pp.
[39] Ueno Y. Kinetics of endogenous CFU-s in mice receiving divided-dose irradiation. *J. Radiat. Res. (Tokyo).* 1975; 16(1): 10-8.
[40] Wilson F.D., Stitzel K.A., Klein A.K., Shifrine M., Graham R., Jones M., Bradley E., Rosenblatt L.S. Quantitative response of bone marrow colony-forming units (CFU-C and PFU-C) in weanling beagles exposed to acute whole-body gamma irradiation. *Radiat. Res.* 1978; 74(2): 289-97.

[41] Lepekhina L.A., Kolesnikova A.I., Konoplyannikov A.G. Radiosensitivity of clonogenic granulocytic macrophage cell precursors in the bone marrow and spleen of tumor-bearing mice. *Radiobiologiia..* 1985; 25(6): 752-5 (Russian).
[42] van Bekkum DW. Radiation sensitivity of the hemopoietic stem cell. *Radiat Res.* 1991; 128(1 Suppl): 4-8.
[43] Meijne E.I., van der Winden-van Groenewegen R.J., Ploemacher R.E., Vos O., David J.A., Huiskamp R. The effects of x-irradiation on hematopoietic stem cell compartments in the mouse. *Exp. Hematol.* 1991; 19(7): 617-23.
[44] Ainsworth E.J., Kelly L.S., Mahlmann L.J., Schooley J.C., Thomas R.H., Howard J., Alpen E.L. Response of colony-forming units-spleen to heavy charged particles. *Radiat Res.* 1983; 96(1): 180-97.
[45] Konoplyannikov A.G., Kolesnikova A.I., Kaplan V.P., Mishanskaia N.I.. Action of neutrons of 2 different energies (0.35 and 0.85 MeV) on mouse bone cells capable of forming granulocyte-macrophage colonies in diffusion chambers. *Radiobiologiia..* 1980; 20(6): 911-3. (Russian).
[46] Thomas F., Gould M.N. Evidence for the repair of potentially lethal damage in irradiated bone marrow. *Radiat .Environ Biophys.* 1982; 20(2): 89-94.
[47] Gan O.I., Todriia T.V. Cellular repair of sublethal radiation damage in 2 subpopulations of the CFUs from embryonal liver and bone marrow of adult mice. *Biull. Eksp. Bio. Med.* 1989; 107(1): 89-91. (Russian).
[48] Cronkite E.P., Inoue T., Bullis J.E. Influence of radiation fractionation on survival of mice and spleen colony-forming units. *Radiat Res.* 1994; 138(2): 266-71.
[49] Sigdestad C.P., Connor A.M., Sims C.S. Modification of neutron-induced hematopoietic effects by chemical radioprotectors. *Int. J. Radiat. Oncol. Biol. Phys.* 1992; 22(4): 807-11.
[50] Patchen ML, MacVittie TJ, Jackson WE. Postirradiation glucan administration enhances the radioprotective effects of WR-2721. *Radiat Res.* 1989; 117(1): 59-69.
[51] Schwartz G.N., Patchen M.L., Neta R., MacVittie T.J. Radioprotection of mice with interleukin-1: relationship to the number of spleen colony-forming units. *Radiat Res.* 1989; 119(1): 101-12.
[52] Konoplyannikov A.G., Konoplyannikova O.A. Radioprotective effect of E. coli endotoxin on hematopoietic stem cells is partially suppressed by inhibiting production of nitric oxide by administering N-omega-nitro-L-arginine. *Radiats. Biol. Radioecol.* 2002; 42(4): 395-8. (Russian).

[53] Uckun F.M., Gillis S., Souza L., Song C.W. Effects of recombinant growth factors on radiation survival on human bone marrow progenitor cells. *Int. J. Radiat. Oncol. Biol. Phys.* 1989; 16(2): 415-35.
[54] Proskuryakov S.Y., Konoplyannikov A.G., Konoplyannikova O.A., Tsyb A.F, Logunov D.Y., Naroditsky B.S., Gintsburg A.L. Effects of gram-positive microorganisms and their products on in vivo survival of hemopoietic clonogenic cells. *Bull. Exp. Biol. Med.* 2008; 145(4): 460-3.
[55] Proskuryakov S.Y., Konoplyannikov A.G., Konoplyannikova O.A, Shevchenko L.I., Verkhovskii Y.G., Tsyb A.F. Possible involvement of NO in the stimulating effect of pifithrins on survival of hemopoietic clonogenic cells. *Biochemistry* (Mosc). 2009; 74(2): 130-6.
[56] Meijne E.I., Ploemacher R.E., Huiskamp R. Sensitivity of murine haemopoietic stem cell populations to X-rays and 1 MeV fission neutrons in vitro and in vivo under hypoxic conditions. *Int. J. Radiat. Bio.l.* 1996; 70(5): 571-7.
[57] Adler S.S., Trobaugh F.E.Jr. Pluripotent (CFU-S) and granulocyte-committed (CFU-C) stem cells in intact and 89Sr marrow-ablated S1/S1d mice. *Cell Tissue Kinet. 1978;* 11(5): 555-66.
[58] Gan O.I., Konoplyannikov A.G. Comparative radiosensitivity of CFUs of the mouse bone marrow and embryonal liver forming 7- and 11-day colonies. *Biull. Eksp. Biol. Med.* 1989; 107(1): 93-5. (Russian).
[59] Konoplyiannikova O.A., Konoplyannikov A.G. Age-related changes in the radiosensitivity of animals and critical cell systems. 1. Survival on irradiation in the "bone marrow" dosage range and the general characteristics of the state of the CFU pool. *Radiobiologiia.* 1977; 17(6): 844-8. (Russian).
[60] Wierenga P.K., Konings A.W. Goralatide (AcSDKP) selectively protects murine hematopoietic progenitors and stem cells against hyperthermic damage. *Exp. Hematol.* 1996; 24(2): 246-52.
[61] Konoplyannikov A.G., Konoplyannikova O.A., Trishkina A.I., Shtein L.V.Radiosensitizing and damaging action of hyperthermia on different biological systems. Radiosensitizing and damaging action of hyperthermia on mouse hematopoietic stem cells. *Radiobiologiia.* 1984; 24(3): 325-9. (Russian).
[62] Wilson F.D., Stitzel K.A., Klein A.K, Shifrine M., Graham R., Jones M., Bradley E., Rosenblatt L.S. Quantitative response of bone marrow colony-forming units (CFU-C and PFU-C) in weanling beagles exposed to acute whole-body gamma irradiation. *Radiat Res.* 1978; 74(2): 289-97.

[63] Klein A.K., Dyck J.A., Shimizu J.A., Stitzel K.A., Wilson F.D., Cain G.R. Effect of continuous, whole-body gamma irradiation upon canine lymphohematopoietic (CFU-GM, CFU-L) progenitors and a possible hematopoietic regulatory population. *Radiat. Res.* 1985; 101(2): 332-50.
[64] Verfaillie C.M. Can human hematopoietic stem cells be cultured ex vivo? *Stem Cells.* 1994; 12(5): 466-76
[65] Kolesnikova A.I., Mishanskaia N.I., Konoplyannikov A.G., Kaplan V.P., Baisogolov G.D. Radiosensitivity of human bone marrow cells forming granulocyte-macrophage colonies in diffusion chambers. *Med. Radiol.* (Mosk). 1982; 27(3): 43-6. (Russian).
[66] Zherbin E.A., Kolesnikova A.I., Konoplyannikov A.G., Khoptynskaia S.K. Radiosensitivity study of human bone marrow cells that form colonies in agar cultures. *Radiobiologiia.* 1978; 18(4): 613-5. (Russian).
[67] Chen B.P., Galy A., Kyoizumi S., Namikawa R., Scarborough J., Webb S., Ford B., Cen D.Z., Chen S.C. Engraftment of human hematopoietic precursor cells with secondary transfer potential in SCID-hu mice. *Blood.* 1994; 84(8): 2497-505.
[68] Hoffman R., Tong J., Brandt J., Traycoff C., Bruno E., McGuire B.W., Gordon M.S., McNiece I., Srour E.F. The in vitro and in vivo effects of stem cell factor on human hematopoiesis. *Stem Cells.* 1993;11(Suppl 2): 76-82.
[69] Majeti R., Park C.Y., Weissman I.L. Identification of a hierarchy of multipotent hematopoietic progenitors in human cord blood. *Cell Stem Cell.* 2007; 1(6): 635-45.
[70] Down J.D., Boudewijn A., van Os R., Thames H.D., Ploemacher R.E. Variations in radiation sensitivity and repair among different hematopoietic stem cell subsets following fractionated irradiation. *Blood.* 1995; 86(1): 122-7.
[71] Abe A., Minami Y., Hayakawa F., Kitamura K., Nomura Y., Murata M., Katsumi A., Kiyoi H., Jamieson C.H., Wang J.Y., Naoe T. Retention but significant reduction of BCR-ABL transcript in hematopoietic stem cells in chronic myelogenous leukemia after imatinib therapy. *Int. J. Hematol.* 2008; 88(5): 471-5.
[72] Carlo-Stella C., Dotti G., Mangoni L., Regazzi E., Garau D., Bonati A., Almici C., Sammarelli G., Savoldo B., Rizzo M.T., Rizzoli V. Selection of myeloid progenitors lacking BCR/ABL mRNA in chronic myelogenous leukemia patients after in vitro treatment with the tyrosine kinase inhibitor genistein. *Blood.* 1996; 88(8): 3091-100.

[73] Abe K., Shimizu R., Pan X., Hamada H., Yoshikawa H., Yamamoto M. Stem cells of GATA1-related leukemia undergo pernicious changes after 5-fluorouracil treatment. *Exp. Hematol.* 2009; 37(4): 435-45.
[74] Botnick L.E., Hannon E.C., Vigneulle R., Hellman S. Differential effects of cytotoxic agents on hematopoietic progenitors. *Cancer Res.* 1981; 41(6): 2338-42.
[75] Stevenson A.F. Haemopoietic recovery during radiation disease: comments on combined-injuries. *Radiat. Environ Biophys.* 1981; 20(1): 29-36.
[76] Walker R.I. Requirement of radioprotectors for military and emergency needs. In: "Perspectives in radioprotection", 1988, Pergamon Press, pp. 13-20.
[77] Kolesnikova A.I., Karpov D.A., Danilova M.A., Pavlov V.V., Kurpeshev O.K., Lephekhina L.A., Kal'sina S.Sh., Konoplyannikov A.G. The effects of whole-body electromagnetic hyperthermia on circulating CFU-GM and on plasma colony-stimulating activity in patients with Hodgkin diseases. *Vopr. Onkol.* 1995; 41(2): 98-100. (Russian).
[78] Konoplyannikova O.A, Konoplyannikov A.G. Radiobiology of intestinal epithelial stem cells. I. Comparative estimation of the radioprotective effect of AET from 4-5 day mortality of mice after radiation and from survival of stem cells of small intestine epithelium. *Radiobiologiia.* 1973; 13(4): 531-6. (Russian).
[79] Konoplyannikov A.G., Konopyiannikova O.A. Radiobiology of intestinal epithelial stem cells. 2. Effect of gamma-radiation dose on 4-5 day mortality of mice and on survival of epithelial stem cells of the small intestine. *Radiobiologiia.* 1973; 13(6): 834-8. (Russian).
[80] Potten C.S. Radiation, the ideal cytotoxic agent for studying the cell biology of tissues such as the small intestine. *Radiat. Res.* 2004; 161(2): 123-36.
[81] Roberts S.A., Hendry J.H., Potten C.S. Intestinal crypt clonogens: a new interpretation of radiation survival curve shape and clonogenic cell number. *Cell Prolif.* 2003; 36(4): 215-31.
[82] Konoplyannikova O.A. Radiobiology of stem cells in intestinal epithelium. Effect of single and multiple preliminary sublethal irradiation of mice on the dose dependence of the survival of stem cells in small intestine epithelium. *Radiobiologiia.* 1988; 28(1): 35-8. (Russian).

[83] Masuda K, Withers HR, Mason KA, Chen KY. Single-dose-response curves of murine gastrointestinal crypt stem cells. *Radiat. Res.* 1977; 69(1): 65-75.

[84] Wambersie A., Dutreix J., Gueulette J., Lellouch J. Early recovery for intestinal stem cells, as a function of dose per fraction, evaluated by survival rate after fractionated irradiation of the abdomen of mice. *Radiat. Res.* 1974; 58(3): 498-515.

[85] Konoplyannikova O.A., Konoplyannikov A.G. Radiobiology of the stem cells of the intestinal epithelium. 3. The effect of irradiation dosage fractionation in 2 age groups of mice. *Radiobiologiia.* 1979; 19(3): 398-401. (Russian).

[86] Huczkowski J., Trott K.R. Dose fractionation effects in low dose rate irradiation of jejunal crypt stem cells. *Int. J. Radiat. Biol. Relat. Stud. Phys. Chem. Med.* 1984; 46(3): 293-8.

[87] Zywietz F., Jung H., Hess A., Franke H.D. Response of mouse intestine to 14 MeV neutrons. *Int. J. Radiat. Biol. Relat. Stud. Phys. Chem. Med.* 1979; 35(1): 63-72.

[88] Hanson W.R., Crouse D.A., Fry R.J., Ainsworth E.J. Relative biological effectiveness measurements using murine lethality and survival of intestinal and hematopoietic stem cells after fermilab neutrons compared to JANUS reactor neutrons and ^{60}Co gamma rays. *Radiat. Res.* 1984 ;100(2): 290-7.

[89] Konoplyannikov A.G. Biological effects of gamma-neutron radiation Cf-252 or fission neutron from the BR-10 reactor on tumor and normal cells and tissues. In: "Californium-252 Isotope for 21st Century Radiotherapy" (Ed. J.G.Wierzbicki), Kluwer Academic Publishers, 1997: 257-61.

[90] Paganetti H., Niemierko A., Ancukiewicz M., Gerweck L.E., Goitein M., Loeffler J.S., Suit H.D. Relative biological effectiveness (RBE) values for proton beam therapy. Int. J. Radiat. *Oncol. Biol. Phys.* 2002; 53(2): 407-21.

[91] Beauduin M., Gueulette J., Coster B.M., Wambersie A. Determination of parameters of the survival curve of the stem cells of the intestinal crypts by LD50 and the regenerated crypt count. Determination of the RBE of p(65) + Be neutrons. *C. R. Seances Soc. Biol .Fil.* 1985; 179(4): 487-92. (French).

[92] Schofield R. The relationship between the spleen colony-forming cell and the haemopoietic stem cell. *Blood Cells.* 1978; 4(1-2): 7–25.

[93] Zhang J., Li L. Stem cell niche: microenvironment and beyond. *J. Biol. Chem.* 2008; 283(15): 9499-503.
[94] Houchen C.W., George R.J., Sturmoski M.A., Cohn S.M. FGF-2 enhances intestinal stem cell survival and its expression is induced after radiation injury. *Am. J. Physiol.* 1999; 276(1 Pt 1): 249-58.
[95] Kim J.A., Kang Y.J., Park G., Kim M., Park Y.O., Kim H., Leem S.H., Chu I.S., Lee J.S., Jho E.H., Oh I.H. Identification of a Stroma-Mediated Wnt/beta-Catenin Signal Promoting Self-Renewal of Hematopoietic Stem Cells in the Stem Cell Niche. *Stem Cells.* 2009; 27(6): 1318-29.
[96] Watson A.J., Pritchard D.M. Lessons from genetically engineered animal models. VII. Apoptosis in intestinal epithelium: lessons from transgenic and knockout mice. *Am. J. Physiol. Gastrointest. Liver Physiol.* 2000; 278(1): G1-5.
[97] Hoyes K.P., Cai W.B., Potten C.S., Hendry J.H. Effect of bcl-2 deficiency on the radiation response of clonogenic cells in small and large intestine, bone marrow and testis. *Int. J. Radiat. Biol.* 2000; 76(11): 1435-42.
[98] Hendry J.H., Cai W.B., Roberts S.A., Potten C.S. p53 deficiency sensitizes clonogenic cells to irradiation in the large but not the small intestine. *Radiat Res.* 1997; 148(3): 254-9.
[99] Merritt A.J., Potten C.S., Kemp C.J., Hickman J.A., Balmain A., Lane D.P., Hall P.A. The role of p53 in spontaneous and radiation-induced apoptosis in the gastrointestinal tract of normal and p53-deficient mice. *Cancer Res.* 1994; 54(3): 614-7.
[100] Mori E., Takahashi A., Yamakawa N., Kirita T., Ohnishi T. High LET heavy ion radiation induces p53-independent apoptosis. *J. Radiat. Res.* (Tokyo). 2009; 50(1): 37-42.
[101] George R.J., Sturmoski M.A., May R., Sureban S.M., Dieckgraefe B.K., Anant S., Houchen C.W. Loss of p21Waf1/Cip1/Sdi1 enhances intestinal stem cell survival following radiation injury. *Am. J. Physiol. Gastrointest. Liver Physiol.* 2009; 296(2): G245-54.
[102] Potten C.S. Stem cells in gastrointestinal epithelium: numbers, characteristics and death. *Philos. Trans. R.. Soc. Lond. B. Biol. Sci.* 1998; 353(1370): 821-30.
[103] Maunda K.K., Moore J.V. Radiobiology and stathmokinetics of intestinal crypts associated with patches of Peyer. *Int. J. Radiat. Biol. Relat. Stud. Phys. Chem. Med.* 1987 ; 51(2): 255-64.
[104] Konoplyannikova OA, Konoplyannikov AG, Vacek A. Radiobiological aspects of increased radioresistance of murine epithelial stem cells from

patches of Peyer. *Radiats. Biol. Radioecol.* 1994; 34(4-5): 514-9. (Russian).
[105] Van Houten N., Blake S.F., Li E.J., Hallam T.A., Chilton D.G., Gourley W.K., Boise L.H., Thompson C.B., Thompson E.B. Elevated expression of Bcl-2 and Bcl-x by intestinal intraepithelial lymphocytes: resistance to apoptosis by glucocorticoids and irradiation. *Int. Immunol.* 1997; 9(7): 945-53.
[106] Potten C.S., Hume W.J., Reid P., Cairns J. The segregation of DNA in epithelial stem cells. *Cell.* 1978; 15(3): 899-906.
[107] Potten C.S., Owen G., Booth D. Intestinal stem cells protect their genome by selective segregation of template DNA strands. *J. Cell. Sci.* 2002; 115(Pt 11): 2381-2388.
[108] Singh V.K., Yadav V.S. Role of cytokines and growth factors in radioprotection. *Exp. Mol. Pathol.* 2005; 78(2): 156–69.
[109] Hancock S.L., Chung R.T., Cox R.S., Kallman R.F. Interleukin 1 beta initially sensitizes and subsequently protects murine intestinal stem cells exposed to photon radiation. *Cancer Res.* 1991; 51(9): 2280–5.
[110] Zaghloul M.S., Dorie M.J., Kallman R.F. Interleukin-1 modulatory effect on the action of chemotherapeutic drugs and localized irradiation of the lip, duodenum, and tumor. *Int. J. Radiat. Oncol. Biol. Phys.* 1993; 26(3): 417–25.
[111] Lamont C., Witsell A., Mauch P.M. Radioprotection of bone marrow stem cell subsets by interleukin-1 and kit-ligand: implications for CFU-S as the responsible target cell population. *Exp. Hematol.* 1997; 25(3): 205-10.
[112] Chen T., Burke K.A., Zhan Y., Wang X., Shibata D., Zhao Y. IL-12 facilitates both the recovery of endogenous hematopoiesis and the engraftment of stem cells after ionizing radiation. *Exp. Hematol.* 2007; 35(2): 203-13.
[113] Neta R., Stiefel S.M., Finkelman F. Herrmann S., Ali N. IL-12 protects bone marrow from and sensitizes intestinal tract to ionizing radiation. *J. Immunol.* 1994; 153(9): 4230-7.
[114] Maj J.G., Paris F., Haimovitz-Friedman A., Venkatraman E., Kolesnick R., Fuks Z. Microvascular function regulates intestinal crypt response to radiation. *Cancer Res.* 2003; 63(15): 4338-41.
[115] Okunieff P., Mester M., Wang J., Maddox T., Gong X., Tang D., Coffee M., Ding I. In vivo radioprotective effects of angiogenic growth factors on the small bowel of C3H mice. *Radiat. Res.* 1998; 150(2): 204-11.

[116] Booth D., Haley J.D., Bruskin A.M., Potten C.S. Transforming growth factor-B3 protects murine small intestinal crypt stem cells and animal survival after irradiation, possibly by reducing stem-cell cycling. *Int. J. Cancer.* 2000; 86(1): 53-9.
[117] Booth D., Potten C.S. Protection against mucosal injury by growth factors and cytokines. *J. Natl. Cancer Inst. Monogr.* 2001;.(29): 16-20.
[118] Farrell C.L., Bready J.V., Rex K.L., Chen J.N., DiPalma C.R., Whitcomb K.L., Yin S., Hill D.C., Wiemann B., Starnes C.O., Havill A.M., Lu Z.N., Aukerman S.L., Pierce G.F., Thomason A., Potten C.S., Ulich T.R., Lacey D.L. Keratinocyte growth factor protects mice from chemotherapy and radiation-induced gastrointestinal injury and mortality. *Cancer Res.* 1998; 58(5): 933-9.
[119] Esworthy R.S., Mann J.R., Sam M., Chu F.F. Low glutathione peroxidase activity in Gpx1 knockout mice protects jejunum crypts from gamma-irradiation damage. *Am. J. Physiol. Gastrointest. Liver Physiol.* 2000; 279(2): G426-36.
[120] Potten C.S., O'Shea J.A., Farrell C.L., Rex K., Booth C. The effects of repeated doses of Keratinocyte growth factor on cell proliferation in the cellular hierarchy of the crypts of the murine small intestine. *Cell Growth Differ.* 2001; 12(5): 265-75.
[121] Khan W.B., Shui C., Ning S., Knox S.J.Enhancement of murine intestinal stem cell survival after irradiation by keratinocyte growth factor. *Radiat. Res.* 1997; 148(3): 248-53.
[122] Leigh B.R., Khan W., Hancock S.L., Knox L. Stem cell factor enhances the survival of murine intestinal stem cells after photon irradiation. *Radiat. Res.* 1995; 142(1): 12-5.
[123] Ainsworth E.J. From endotoxins to newer immunomodulators: survival-promoting effects of microbial polysaccharide complexes in irradiated animals. *Pharmacol. Ther.* 1988; 39(1-3): 223-41.
[124] Riehl T., Cohn S., Tessner T., Schloemann S., Stenson W.F.Lipopolysaccharide is radioprotective in the mouse intestine through a prostaglandin-mediated mechanism. *Gastroenterology.* 2000; 118(6): 1106-16.
[125] Anant S., Murmu N., Houchen C.W., Mukhopadhyay D., Riehl T.E., Young S.G., Morrison A.R., Stenson W.F., Davidson N.O. Apobec-1 protects intestine from radiation injury through posttranscriptional regulation of cyclooxygenase-2 expression. *Gastroenterology.* 2004 ; 127(4): 1139-49.

[126] Riehl T.E., Newberry R.D., Lorenz R.G., Stenson W.F. TNFR1 mediates the radioprotective effects of lipopolysaccharide in the mouse intestine. *Am. J. Physiol. Gastrointest. Liver Physiol.* 2004; 286(1): G166-73.

[127] Hassan F., Islam S., Mu M.M., Ito H., Koide N., Mori I., Yoshida T., Yokochi T. Lipopolysaccharide prevents doxorubicin-induced apoptosis in RAW 264.7 macrophage cells by inhibiting p53 activation. *Mol. Cancer Res.* 2005; 3(7): 373-9.

[128] DeGowin R.L., Fisher P.G., An D. Differential elaboration of prostaglandin E2 by cells of the hemopoietic microenvironment in response to endotoxin. *J. Lab. Clin Med.* 1987; 109(6): 679-86.

[129] Ruifrok A.C., Mason K.A., Thames H.D. Changes in clonogen number and radiation sensitivity in mouse jejunal crypts after treatment with dimethylsulfoxide and retinoic acid. *Radiat. Res.* 1996; 145(6): 740-5.

[130] Lehnert S. Radioprotection of mouse intestine by inhibitors of cyclic AMP phosphodiesterase. *Int. J. Radiat. Oncol. Biol. Phys.* 1979; 5(6): 825-33.

[131] Murray D., Altschuler E.M., Hunter N., Milas L. Protection by WR-3689 against gamma-ray-induced intestinal damage: comparative effect on clonogenic cell survival, mouse survival, and DNA damage. *Radiat. Res.* 1989; 120(2): 339-51.

[132] Ramdas J., Warrier R.P., Scher C., Larussa V. Effects of amifostine on clonogenic mesenchymal progenitors and hematopoietic progenitors exposed to radiation. *J. Pediatr. Hematol. Oncol.* 2003; 25(1): 19-26.

[133] Cohn S.M., Schloemann S., Tessner T., Seibert K., Stenson W.F. Crypt stem cell survival in the mouse intestinal epithelium is regulated by prostaglandins synthesized through cyclooxygenase-1. *J. Clin. Invest.* 1997; 99(6): 1367-79.

[134] Riehl T.E., George R.J., Sturmoski M.A., May R., Dieckgraefe B., Anant S., Houchen C.W. Azoymethane protects intestinal stem cells and reduces crypt epithelial mitosis through a COX-1-dependent mechanism. *Am. J. Physiol. Gastrointest. Liver Physiol.* 2006; 291(6): G1062- 70.

[135] Proskuryakov S.Y., Konoplyannikov A.G., Konoplyannikova O.A., Ulyanova L.P., Tsyb A.F. Role of cyclooxygenases in the stimulatory effect of carcinogen 1,2-dimethylhydrazine on stem cell survival in the intestinal epithelium and bone marrow. *Bull. Exp. Biol. Med.* 2008; 146(4): 540-2.

[136] Houchen C.W., Stenson W.F., Cohn S.M. Disruption of cyclooxygenase-1 gene results in an impaired response to radiation

injury. *Am. J. Physiol. Gastrointest. Liver Physiol.* 2000 ; 279(5): G858-65.
[137] Komarov P.G., Komarova E.A., Kondratov R.V., Christov-Tselkov K., Coon J.S., Chernov M.V., Gudkov A.V. A chemical inhibitor of p53 that protects mice from the side effects of cancer therapy. *Science.* 1999; 285(5434): 1733-7.
[138] Singh S.K., Clarke I.D., Hide T., Dirks P.B. Cancer stem cells in nervous system tumors. *Oncogene.* 2004; 23(43): 7267-73.
[139] Baumann M., Krause M., Hill R. Exploring the role of cancer stem cells in radioresistance. *Nat. Rev. Cancer.* 2008; 8(7): 545-54.
[140] Hambardzumyan D., Becher. O.J., Holland E.C. Cancer stem cells and survival pathways. *Cell Cycle.* 2008; 7(10): 1371-8.
[141] Zherbin E.A., Kolesnikova A.I., Konoplyannikov A.G., Khoptynskaia S.K. Radiosensitivity of human bone marrow cells that form fibroblast colonies in monolayer cultures. *Med. Radiol.* (Mosk). 1978; 23(7): 48-51. (Russian).
[142] Zherbin E.A., Kolesnikova A.I., Konoplyannikov A.G., Khoptynskaia S.K., Kapchigashev S.P. Action of gamma rays and fast neutrons on monolayer human bone marrow cultures. *Med. Radiol.* (Mosk). 1978; 23(8): 53-7. (Russian).
[143] Kolesnikova A.I., Konoplyannikov A.G., Hendry J.H. Differential sensitivity of two predominant stromal progenitor cell subpopulations in bone marrow to single and fractionated radiation doses. *Radiat Res.* 1995; 144(3): 342-5.
[144] Wang S.B., Hendry J.H., Testa N.G. Sensitivity and recovery of stromal progenitor cells (CFU-F) in mouse bone marrow given gamma-irradiation at 0.65 Gy per day. *Biomed Pharmacother.* 1987; 41(1): 48-50.
[145] Bajsogolov G.D., Siskin I.P., Choptynskaja S.K., Kolesnikova A.I., Misanskaja N.I. Late effects of radiation. The condition of the stroma in irradiated and intact areas of human bone marrow. *Radiobiol. Radiother.* (Berl). 1982; 23(1): 31-5. (German).
[146] Konoplyannikov A.G., Waïnson A.A., Kolesnikova A.I., Zaïtsev A.V., Kal'sina S.Sh., Lepekhina L.A.The radioprotective effect of hypoxia on clonogenic cells of rat bone marrow stroma (CFU-F). *Radiobiologiia.* 1992; 32(5): 720-4. (Russian).
[147] Ma J., Shi M, Li J., Chen B., Wang H., Li B., Hu J., Cao Y., Fang B., Zhao R.C. Senescence-unrelated impediment of osteogenesis from Flk1+ bone marrow mesenchymal stem cells induced by total body

irradiation and its contribution to long-term bone and hematopoietic injury. *Haematologica.* 2007; 92(7): 889-96.
[148] Wang Y., Schulte B.A., Zhou D. Hematopoietic stem cell senescence and long-term bone marrow injury. *Cell Cycle.* 2006; 5(1): 35-8.
[149] Semenkova I.V., Kolesnikova A.I., Lepechina L.A., Kal'sina S.Sh., Agaeva E.V., Konoplyannikov A.G. Differences in rat or human bone marrow mesenchymal stem cells and produced from them cardiomyoblasts. *Cytokines and Inflammation..* 2005; 4(2): 112 (Russian).
[150] Uchibori R., Okada T., Ito T., Urabe M., Mizukami H., Kume A., Ozawa K. Retroviral vector-producing mesenchymal stem cells for targeted suicide cancer gene therapy. *J. Gene Med.* 2009; 11(5): 373-81.
[151] Aluigi M., Fogli M., Curti A., Isidori A., Gruppioni E., Chiodoni C., Colombo M.P., Versura P., D'Errico-Grigioni A., Ferri E., Baccarani M., Lemoli R.M. Nucleofection is an efficient nonviral transfection technique for human bone marrow-derived mesenchymal stem cells. *Stem Cells.* 2006; 24(2): 454-61.
[152] Meistrich M.L., Finch M., Lu C.C., de Ruiter-Bootsma A.L., de Rooij D.G., Davids J.A. Strain differences in the response of mouse testicular stem cells to fractionated radiation. *Radiat. Res.* 1984; 97(3): 478-87.
[153] de Ruiter-Bootsma A.L., Kramer M.F., de Rooij D.G., Davis J.A. Survival of spermatogonial stem cells in the mouse after split-dose irradiation with fission neutrons of 1-MeV mean energy. Effect of the fractionation interval. *Radiat. Res.* 1979; 79(2): 289-97.
[154] Suzuki N., Withers H.R., Hunter N. Radiosensitization of mouse spermatogenic stem cells by Ro-07-0582. *Radiat. Res.* 1977; 69(3): 598-601.
[155] Suzuki N., Withers H.R. Exponential decrease during aging and random lifetime of mouse spermatogonial stem cells. *Science.* 1978; 202(4373): 1214-5.
[156] Reid BO, Mason KA, Withers HR, West J. Effects of hyperthermia and radiation on mouse testis stem cells. *Cancer Res.* 1981; 41(11 Pt1): 4453-7.
[157] Kostereva N., Hofmann M.C Regulation of the spermatogonial stem cell niche. *Reprod. Domest. Anim..* 2008; 43 Suppl 2: 386-92.
[158] Baumann M., Krause M., Thames H., Trott K., Zips D. Cancer stem cells and radiotherapy. *Int. J. Radiat. Biol.* 2009; 85(5): 391-402.

[159] Kang M.K., Hur B.I., Ko M.H., Kim C.H., Cha S.H., Kang S.K. Potential identity of multi-potential cancer stem-like subpopulation after radiation of cultured brain glioma. *BMC Neurosci.* 2008; 9:15.
[160] Ischenko I., Seeliger H., Schaffer M., Jauch K.W., Bruns C.J. Cancer stem cells: how can we target them? *Curr. Med .Chem.* 2008; 15(30): 3171-84.
[161] Konoplynnikov A.G. Cell basis of human radiation effects. In: "Radiatiom medicine, v.1", 2005; M., Izdat: 189-277.
[162] Cairnie A.B., Millen B.H. Fission of crypts in the small intestine of the irradiated mouse. *Cell Tissue Kinet.* 1975; 8(2): 189-96.
[163] Ziablitskiĭ V.M., Konoplyannikov A.G., Maslennikova R.L., Romanovskaia V.N. Quantitative evaluation of hematopoietic stem cell migration. *Radiobiologiia.* 1980 ; 20(3): 368-72. (Russian).
[164] Schulz C., von Andrian U.H., Massberg S. Hematopoietic stem and progenitor cells: their mobilization and homing to bone marrow and peripheral tissue. *Immunol. Res.* 2009; 44(1-3): 160-8.
[165] Campbell F., Williams G.T., Appleton M.A., Dixon M.F., Harris M., Williams E.D. Post-irradiation somatic mutation and clonal stabilisation time in the human colon. *Gut.* 1996 ; 39(4): 569-73.
[166] Lotem J., Sachs L. Epigenetics and the plasticity of differentiation in normal and cancer stem cells. *Oncogene.* 2006; 25(59): 7663-72.
[167] Overgaard J, Overgaard M. Hyperthermia as an adjuvant to radiotherapy in the treatment of malignant melanoma. *Int. J. Hyperthermia.* 1987; 3(6): 483-501.
[168] Konoplyannikov A.G. Thermoradiotherapy of tumors in the USSR. In: "Soviet Medical Reviews/Section F; Oncology Reviews", 1991; 3(5): 67-112.
[169] Hall EJ, Roizin-Towle L. Biological effects of heat. *Cancer Res.* 1984; 44(10 Suppl): 4708s-13s.
[170] Engin K. Biological rationale for hyperthermia in cancer treatment (II). *Neoplasma.* 1994; 41(5): 277-83.
[171] Hahn G.M., Li G.C. Interactions of hyperthermia and drugs: treatments and probes. *Natl. Cancer Inst. Monogr.* 1982; 61: 317-23.
[172] Bettaieb A., Averill-Bates D.A. Thermotolerance induced at a fever temperature of 40 degrees C protects cells against hyperthermia-induced apoptosis mediated by death receptor signalling. *Biochem. Cell Biol.* 2008; 86(6): 521-38.
[173] Takahashi A., Yamakawa N., Mori E., Ohnishi K., Yokota S., Sugo N., Aratani Y., Koyama H., Ohnishi T. Development of thermotolerance

requires interaction between polymerase-beta and heat shock proteins. *Cancer Sci.* 2008; 99(5): 973-8.
[174] Horowitz M., Robinson S.D. Heat shock proteins and the heat shock response during hyperthermia and its modulation by altered physiological conditions. *Prog. Brain Res.* 2007; 162: 433-46.
[175] Kozanoglu I., Boga C., Ozdogu H., Maytalman E., Ovali E., Sozer O. A detachment technique based on the thermophysiologic responses of cultured mesenchymal cells exposed to cold. *Cytotherapy.* 2008; 10(7): 686-9.
[176] Kocabiyik S. Essential structural and functional features of small heat shock proteins in molecular chaperoning process. *Protein Pept. Lett.* 2009; 16(6): 613-22.
[177] Konoplyannikov A.G., Konoplyannikova O.A., Trishkina A.I., Shteĭn L.V. Radiosensitizing and damaging action of hyperthermia on different biological systems. Radiosensitizing and damaging action of hyperthermia on mouse hematopoietic stem cells. *Radiobiologiia.* 1984; 24(3): 325-9. (Russian).
[178] Dewey W.C. Arrhenius relationships from the molecule and cell to the clinic. *Int. J. Hyperthermia.* 2009; 25(1): 3-20.
[179] Shteĭn L.V., Konoplyannikov A.G. Radiosensitizing and damaging effect of hyperthermia on different biological systems. Radiosensitizing and damaging effect of hyperthermia on cells of mouse La leukemia. *Radiobiologiia.* 1983; 23(4): 489-92. (Russian).
[180] Synzynys B.I., Kolesnikova A.I., Konoplyannikova A.G. Study of DNA synthesis in short-term cultures of mammalian cells in the evaluation of the reaction of tumor and normal tissues to irradiation and hyperthermia. *Radiobiologiia.* 1985; 25(2): 179-84. (Russian).
[181] Kaplan V.P., Chernysheva N.M., Konoplyannikov A.G. Effect of radiation, thermal or combined radiation-thermal damage to a population of progenitor cells of the granulocyte-macrophage series in mouse bone marrow. *Radiobiologiia.* 1987; 27(6): 753-6. (Russian).
[182] Kolesnikova A.I., Kal'sina S.Sh., Lepekhina L.A., Shteĭn L.V., Grigor'ev A.N., Kurpeshev O.K., Semichastnova L.M., Konoplyannikov A.G. Thermosensitivity of clonogenic cells and the induction of thermal tolerance. *Med. Radiol.* (Mosk). 1987 ; 32(1): 67-9. (Russian).
[183] Lebedeva T.V., Konoplyannikova O.A. Radiosensitizing effect of hyperthermia (41.5^0 C, 30 min) in stem cells of mouse spermatogenic epithelium. *Radiats Biol Radioecol.* 1993; 33(4): 564-6. (Russian).

[184] Konoplyannikov A.G. Current problems of thermobiology. *Med. Radiol.* (Mosk). 1987; 32(1): 53-6. (Russian).
[185] Symonds RP, Wheldon TE, Clarke B, Bailey G. A comparison of the response to hyperthermia of murine haemopoietic stem cells (CFU-S) and L1210 leukaemia cells: enhanced killing of leukaemic cells in presence of normal marrow cells. *Br. J. Cancer.* 1981; 44(5): 682-91.
[186] Setroikromo R., Wierenga P.K., van Waarde M.A., Brunsting J.F., Vellenga E., Kampinga H.H. Heat shock proteins and Bcl-2 expression and function in relation to the differential hyperthermic sensitivity between leukemic and normal hematopoietic cells. *Cell Stress Chaperones.* 2007; 12(4): 320-30.
[187] Leith J.T., DeWyngaert K., Dexter D.L., Calabresi P., Glicksman A.S. Differential sensivity of three adenocarcinoma lines to hyperthermic cell killing. *Natl. Cancer Inst. Monogr.* 1982; 61: 381-3.
[188] Mimeault M., Hauke R., Batra S.K. Stem cells: a revolution in therapeutics-recent advances in stem cell biology and their therapeutic applications in regenerative medicine and cancer therapies. *Clin Pharmacol Ther.* 2007; 82(3): 252-64.
[189] Liang H., Zhan H.J., Wang B.G., Pan Y., Hao X.S. Change in expression of apoptosis genes after hyperthermia, chemotherapy and radiotherapy in human colon cancer transplanted into nude mice. *World J. Gastroenterol.* 2007; 13(32): 4365-71.
[190] Proskuriakov S.Ia., Konoplyannikov A.G., Gabaĭ V.L. Cellular necrosis in genesis and therapy of diseases. *Ter. Arkh..* 2006; 78(1): 65-9.(Russian).
[191] Proskuryakov S.Y., Konoplyannikov A.G., Gabai V.L. Necrosis: a specific form of programmed cell death? *Exp Cell Res.* 2003; 283(1): 1-16.
[192] Timiryasova T.M., Gridley D.S., Chen B., Andres M.L., Dutta-Roy R., Miller G., Bayeta E.J., Fodor I. Radiation enhances the anti-tumor effects of vaccinia-p53 gene therapy in glioma. *Technol. Cancer Res. Treat.* 2003; 2(3): 223-35.
[193] Takahashi A., Ohnishi K., Ota I., Asakawa I., Tamamoto T., Furusawa Y., Matsumoto H., Ohnishi T. p53-dependent thermal enhancement of cellular sensitivity in human squamous cell carcinomas in relation to LET. *Int. J. Radiat. Biol.* 2001; 77(10): 1043-51.
[194] Tamamoto T., Yoshimura H., Takahashi A., Asakawa I., Ota I., Nakagawa H., Ohnishi K., Ohishi H., Ohnishi T. Heat-induced growth inhibition and apoptosis in transplanted human head and neck squamous

cell carcinomas with different status of p53. *Int. J. Hyperthermia.* 2003; 19(6): 590-7.

[195] Tokalov S.V., Pieck S., Gutzeit H.O. Varying responses of human cells with discrepant p53 activity to ionizing radiation and heat shock exposure. *Cell Prolif.* 2007; 40(1): 24-37.

[196] Guan J., Stavridi E., Leeper D.B., Iliakis G. Effects of hyperthermia on p53 protein expression and activity. *J Cell Physiol.* 2002; 190(3): 365-74.

[197] Konoplyannikov A.G., Konoplyiannikova O.A., Proskuriakov S.Ia. "Ischemia/reperfusion" for stem cells of two "critical" cell renewal systems of organism. *Radiats. Biol. Radioecol.* 2005 ; 45(5): 605-9. (Russian).

[198] Saikumar P., Dong Z., Weinberg J.M., Venkatachalam M.A. Mechanisms of cell death in hypoxia/reoxygenation injury. *Oncogene.* 1998; 17(25): 3341-9.

[199] Brunelle J.K., Chandel N.S. Oxygen deprivation induced cell death: an update. *Apoptosis.* 2002; 7(6): 475-82.

[200] Saikumar P., Venkatachalam M.A. Role of apoptosis in hypoxic/ischemic damage in the kidney. *Semin. Nephrol.* 2003; 23(6): 511-21.

[201] Tang J., Xie Q., Pan G., Wang J., Wang M. Mesenchymal stem cells participate in angiogenesis and improve heart function in rat model of myocardial ischemia with reperfusion. *Eur. J. Cardiothorac. Surg.* 2006; 30(2): 353-61.

[202] Tsyb A.F., Roshal' L.M., Yuzhakov V.V., Konoplyannikov A.G., Sushkevich G.N., Bandurko L.N., Ingel' I.E., Semenova Zh.B., Konoplyannikova O.A., Lepekhina L.A., Kal'sina S.Sh., Verkhovskii Y.G., Shevchuk A.S., Semenkova I.V. Morphofunctional study of the therapeutic effect of autologous mesenchymal stem cells in experimental diffuse brain injury in rats. *Bull. Exp. Biol. Med.* 2006; 142(1): 140-7.

[203] He A., Jiang Y., Gui C., Sun Y., Li J., Wang J.A. The antiapoptotic effect of mesenchymal stem cell transplantation on ischemic myocardium is enhanced by anoxic preconditioning. *Can. J. Cardiol.* 2009; 25(6): 353-8.

[204] Shashkov V.S., Anashkin O.D., Suvorov N.N., Manaeva I.A. Effectiveness of serotonin, mexamine, AET and cystamine during multiple administration following gamma-irradiation. *Radiobiologiia.* 1971; 11(4): 621-3. (Russian).

[205] Maisin JR. Chemical protection against ionizing radiation. *Adv. Space Res.* 1989; (10): 205-12.
[206] Yang S.L., Chen L.J., Kong Y., Xu D., Lou Y.J. Sodium nitroprusside regulates mRNA expressions of LTC4 synthesis enzymes in hepatic ischemia/reperfusion injury rats via NF-kappaB signaling pathway. *Pharmacology.* 2007; 80(1): 11-20.
[207] Malkina R.M. Survival of irradiated animals and the preservation of CFUs after the administration of various radioprotectors. *Radiobiologiia.* 1984; 24(5): 651-4. (Russian).
[208] Trishkina A.I., Konoplyiannikov A.G. The radiosensitivity of hematopoietic stem cells from mice forming splenic colonies after 8 and 12 days following bone marrow cell transplantation (CFU-S-8 and CFU-S-12). *Radiobiologiia.* 1992; 32(2): 207-10. (Russian).
[209] Trishkina A.I., Konoplyannikov A.G. The methods of irradiation dose fractionation and rate change used in studying the capacity for the postradiation repair of CFU-s forming splenic colonies after 8 and 12 days following bone marrow cell transplantation. *Radiobiologiia.* 1992; 32(2): 312-6. (Russian).
[210] Asegawa A.T., Landahl H.D. Studies on spleen oxygen tension and radioprotection in mice with hypoxia, serotonin, and p-aminopropiophenone. *Radiat Res.* 1967; 31(3): 389-99.
[211] Programmed cell death. (Ed. Novikov V.S.). Nauka, St.-Petersburg, 1996, 278 pp. (Russian).
[212] Ringe J., Strassburg S., Neumann K., Endres M., Notter M., Burmester G.R., Kaps C., Sittinger M. Towards in situ tissue repair: human mesenchymal stem cells express chemokine receptors CXCR1, CXCR2 and CCR2, and migrate upon stimulation with CXCL8 but not CCL2. *J. Cell Biochem.* 2007; 101(1): 135-46.
[213] Rosová I., Dao M., Capoccia B., Link D., Nolta J.A. Hypoxic preconditioning results in increased motility and improved therapeutic potential of human mesenchymal stem cells. *Stem Cells.* 2008 ; 26(8): 2173-82.
[214] Wisel S., Khan M., Kuppusamy M. L., Mohan I. K., Chacko S.M., Rivera B. K., Sun B. C., Hideg K., Kuppusamy P. Pharmacological preconditioning of mesenchymal stem cells with trimetazidine (1-[2,3,4-trimethoxybenzyl]piperazine) protects hypoxic cells against oxidative stress and enhances recovery of myocardial function in infarcted heart through Bcl-2 expression. *J. Pharmacol. Exp. Ther.* 2009; 329(2): 543-550.

INDEX

A

abdomen, 33
acid, 12, 14, 37
activation, 20, 21, 23, 37
activation energy, 20
actuators, 21
acute, 3, 17, 23, 29, 31
adenocarcinoma, 21, 42
adipocytes, 16
administration, 15, 17, 24, 30, 43, 44
adult, vii, viii, 1, 2, 3, 8, 10, 11, 13, 14, 15, 16, 17, 19, 23, 25, 28, 30
adult stem cells, vii, viii, 1, 2, 3, 8, 11, 13, 14, 15, 16, 17, 19, 23, 25
adult tissues, 1
agar, 32
age, 2, 16, 33
agent, viii, 16, 24, 33
agents, vii, viii, 1, 2, 8, 13, 14, 15, 17, 19, 23, 32
aging, 16, 39
alpha, 13
angiogenesis, 43
angiogenic, 36
animal models, 34
animals, viii, 8, 11, 12, 14, 15, 17, 19, 24, 31, 37, 44

anoxic, 43
antiapoptotic, 43
Antibodies, 12
anticancer, 22
anti-inflammatory agents, 14
anti-tumor, 42
apoptosis, 10, 14, 17, 21, 23, 25, 34, 35, 37, 41, 42, 43
apoptotic, 10, 12, 14, 15, 21
application, 12, 15, 19, 22
arginine, 30
arrest, 10, 14
autologous bone, 21, 23

B

bacteria, 13
bacterial, 12
Bax, 22
Bcl-2, 21, 22, 35, 42, 44
binding, 13
biological activity, 9
biological systems, 31, 41
blood, 32
bone marrow, vii, 1, 3, 7, 15, 17, 21, 23, 28, 29, 30, 31, 32, 34, 36, 38, 39, 40, 41, 44
bone marrow transplant, 7
bowel, 36
brain, 40, 43

brain injury, 43

C

caffeine, 14
cancer, viii, 8, 11, 15, 17, 19, 26, 27, 38, 39, 40, 42
cancer cells, 26
cancer stem cells, viii, 12, 15, 17, 38, 40
cancer treatment, 19, 40
carcinogen, 15, 17, 38
carcinogenesis, 11
carcinogens, viii, 14, 17, 26
carcinomas, 42, 43
cardiomyocytes, 16
cell, vii, viii, 1, 2, 3, 4, 5, 6, 7, 8, 9, 10, 11, 13, 14, 15, 16, 17, 19, 22, 23, 25, 27, 29, 31, 32, 33, 34, 35, 36, 37, 38, 39, 40, 41, 42, 43, 44
cell culture, 3, 8, 15, 16
cell cycle, 2, 10, 13
cell death, vii, 11, 19, 25, 42, 43, 44
cell differentiation, 14
cell division, 1, 11
cell killing, 42
cell line, 2, 7, 16, 21, 27, 29
cell transplantation, 44
central nervous system, 29
Chaperones, 42
charged particle, 7, 30
chemical agents, 8
chemical reactions, 20
chemicals, 2
chemokine, 44
chemokine receptor, 44
chemotherapeutic drugs, 35
chemotherapy, 8, 17, 19, 26, 36, 42
chondrocytes, 16
chronic myelogenous, 32
classical, vii
colitis, 13
colon, 21, 40, 42

colon cancer, 42
components, 6, 13, 17, 23
concentration, 14
Congress, iv
control, 10, 14, 20, 24
control group, 14, 24
COX-1, 38
crypt stem cells, 33, 36
culture, 7, 25
cyclic AMP, 37
cycling, 36
cyclooxygenase, 13, 37, 38
cyclooxygenase-2, 37
cyclooxygenases, 14, 38
cytokine, 12
cytokines, 10, 11, 35, 36
cytotoxic, 14, 32, 33
Cytotoxic, 29
cytotoxic action, 14
cytotoxic agents, 32

D

death, vii, 2, 4, 5, 8, 9, 12, 14, 17, 21, 25, 35, 41
deficiency, 34
deficit, 3, 14
delivery, 16, 23
deprivation, 43
destruction, 10, 12, 14
detachment, 41
differentiation, 2, 11, 14, 16, 17, 20, 40
diffusion, 30, 31
dimensionality, 6
dimethylsulfoxide, 14, 37
direct action, 12
direct measure, 28
disorder, 21
distribution, 5, 9, 20
division, 1, 11
DNA polymerase, 19
DNA repair, 10, 21

Index

donor, 7, 24
donors, 24
dosage, viii, 31, 33
dream, 27
drugs, 17, 19, 26, 35, 40
duodenum, 35
duration, 19, 25

E

E. coli, 30
elaboration, 37
electromagnetic, 33
electron, 16
embryonic stem, 2
embryonic stem cells, 2, 28
endothelial cell, 12
endothelial cells, 12
endothelium, 1, 12
endotoxins, 37
energy, 6, 39
energy transfer, 6
engraftment, 36
enzymes, 17, 21, 44
Epi, 40
epithelia, vii, 15
epithelial cell, 13, 28
epithelial cells, 13, 28
epithelial stem cell, 33, 35
epithelium, vii, viii, 1, 2, 3, 9, 11, 14, 15, 23, 29, 33, 34, 35, 37, 38, 42
erythroid, 7
exponential functions, 7
exposure, 43
extrapolation, 4, 5, 7, 8, 9

F

family, 10, 13, 17, 21
family members, 21
fetal, 8
fever, 41

FGF-2, 10, 12, 34
fibroblast, 13, 28, 38
fibroblast growth factor, 13
fibroblasts, 1, 10, 13
fission, 17, 31, 34, 39
Ford, 32
fractionation, 10, 30, 33, 39, 44

G

gamma radiation, 28
gamma rays, 34, 38
gamma-ray, 37
gastrointestinal, 33, 35, 36
gastrointestinal tract, 35
gene, 10, 15, 16, 21, 38, 39, 42
gene therapy, 16, 39, 42
generation, 12
genes, 10, 14, 17, 42
genistein, 32
genome, 35
genotype, 14
glial, 29
glioma, 40, 42
glucocorticoids, 35
glutathione, 13, 36
glutathione peroxidase, 36
goblet cells, 2
gram-negative bacteria, 13
granulocyte, 30, 31, 41
graph, 4
greek, 6
groups, 1, 14, 24, 33
growth, 8, 10, 11, 24, 30, 35, 36, 43
growth factor, 10, 11, 30, 35, 36
growth factors, 10, 11, 30, 35, 36
growth inhibition, 43
gut, 27

H

heart, 23, 43, 44

heat, viii, 19, 22, 40, 41, 43
heat shock protein, viii, 19, 41
heating, 19
helper cells, 12
hematopoiesis, 8, 11, 17, 32, 36
hematopoietic, vii, viii, 2, 3, 7, 8, 12, 17, 25, 27, 28, 29, 30, 31, 32, 34, 37, 39, 40, 41, 42, 44
hematopoietic cells, 42
hematopoietic progenitor cells, 12
Hematopoietic stem and progenitor cells, 40
hematopoietic stem cell, vii, viii, 3, 7, 8, 28, 29, 30, 31, 32, 34, 40, 41, 44
Hematopoietic stem cell, 39
hematopoietic stem cells, vii, viii, 3, 7, 8, 28, 30, 31, 32, 34, 41, 44
hematopoietic system, 2, 17
hepatocytes, 29
histological, 9
Holland, 38
homeostasis, 8, 10, 11, 19
HR, 33, 40
HSC, 3, 4, 7, 8, 10, 12, 13, 15, 17, 20, 24
human, 3, 8, 12, 15, 16, 21, 28, 30, 31, 32, 38, 39, 40, 42, 43, 44
human embryonic stem cells, 28
human mesenchymal stem cells, 28, 44
humans, 8, 19
hyperthermia, vii, viii, 19, 22, 31, 33, 40, 41, 42, 43
hypothesis, 15, 24
hypoxia, 16, 17, 23, 39, 43, 44
hypoxic, 24, 31, 43, 44
hypoxic cells, 44
injuries, 27, 32
injury, iv, 34, 35, 36, 37, 38, 39, 43, 44
interaction, 17, 41
interactions, 10
interleukin, 30, 36
interleukin-1, 30, 36
Interleukin-1, 12, 35
interval, 39

intestinal tract, 36
intestine, 9, 10, 11, 25, 33, 37
ionizing radiation, vii, viii, 1, 4, 5, 7, 16, 17, 19, 23, 29, 36, 43
irradiation, 1, 2, 3, 4, 6, 7, 9, 10, 12, 13, 14, 15, 16, 17, 20, 21, 24, 29, 31, 32, 33, 34, 35, 36, 37, 38, 39, 40, 41, 43, 44
ischemia, vii, viii, 23, 25, 44
ischemic, vii, viii, 43
Islam, 37
Isotope, 34

J

jejunum, 36
Jung, 33

K

keratinocyte, 36
keratinocytes, 13
kidney, 43
kidneys, 1, 23
killing, 42
kinase, 32
kinetics, 20
knockout, 10, 34, 36

L

labeling, 21
large intestine, 2, 10, 11, 34
laws, 4
leukaemia, 42
leukemia, 32, 41
leukemic, 21, 42
leukemic cells, 21
lifespan, 10, 12
lifetime, 39
ligand, 36
linear, 4, 5, 6, 8, 9, 15
lipopolysaccharide, 37

lipopolysaccharides, 13
liver, 1, 30, 31
local action, 21
London, 28, 29
LPS, 13
lungs, 1
lymphocytes, 12, 35
lymphoma, 16

M

macrophage, 29, 30, 31, 37, 41
macrophages, 7, 12, 21
magnetic, iv
maintenance, 17
malignant, viii, 40
malignant melanoma, 40
malignant tumors, viii
mammalian cell, 7, 13, 20, 29, 41
mammalian cells, 7, 13, 20, 29, 41
marrow, 3, 7, 15, 21, 23, 31, 42
Mars, 28
mathematics, 9
measurement, 4, 28
medicine, 40
mesenchymal stem cell, vii, 3, 15, 21, 39, 43, 44
mesenchymal stem cells, vii, 3, 15, 21, 28, 39, 43, 44
metastases, 12
mice, 7, 8, 10, 12, 14, 16, 17, 20, 24, 29, 30, 31, 32, 33, 34, 35, 36, 38, 42, 44
microbial, 37
microenvironment, 10, 14, 18, 28, 34, 37
microorganisms, 30
microwave, 21
microwave radiation, 21
migration, 17, 21, 23, 40
military, 32
mitosis, 38
mitotic, 17
models, 10, 13, 17, 34

modulation, 41
molecular biology, 8
molecular mechanisms, vii, 27
monocytes, 12
monolayer, 28, 38
morphological, 20
mortality, 33, 36
mouse, 28, 29, 30, 31, 33, 37, 38, 39, 40, 41, 42
mRNA, 32, 44
MSC, 3, 15, 16, 17, 21, 23, 25
mucosa, 12, 28
multipotent, 32
muscle, 1
mutant, 14, 22
mutant cells, 22
mutation, 40
mutations, 18
myeloid, 21, 32
myocardial ischemia, 43
myocardial regeneration, 27
myocardium, 43

N

neck, 43
necrosis, 13, 21, 23, 25, 42
nerve, 1
nervous system, 23, 38
neutrons, 7, 9, 30, 31, 33, 34, 38, 39
New Jersey, 27
New York, iii, iv
nitric oxide, 30
normal, 1, 19, 21, 28, 34, 35, 40, 41, 42
normal conditions, 1

O

Oncogene, 27, 38, 40, 43
Oncology, 40
organism, vii, viii, 1, 3, 8, 16, 17, 21, 23, 25, 43

osteoblasts, 16
oxidative, 44
oxidative stress, 44
oxide, 30
oxygen, vii, viii, 16, 23, 25, 44
Oxygen, 43

P

p53, 10, 14, 15, 17, 21, 34, 35, 37, 38, 42, 43
parameter, 5, 8, 20
parenchymal, 1, 23, 25, 29
parenchymal cell, 1, 23, 25
Paris, 29, 36
PARP, 10
PARP-1, 10
particles, 7, 30
pathogenesis, 18, 27
pathways, vii, 11, 38
patients, 15, 17, 32, 33
pH, 19
phenomenology, vii
phosphodiesterase, 14, 37
photon, 35, 37
physiological, 1, 11, 16, 21, 23, 41
pig, 28
plasma, 33
plastic, 23
plasticity, 2, 18, 40
play, 10, 15, 17, 23
pluripotency, 27
Poisson, 5, 9
Poisson distribution, 9
polymerase, 21, 41
polypeptides, 12
polysaccharide, 37
population, 2, 3, 6, 10, 13, 15, 17, 31, 36, 41
preconditioning, 25, 43, 44
precursor cells, 15, 32
probability, 9, 11, 24
production, viii, 12, 14, 17, 30

progenitor cells, 4, 7, 8, 16, 21, 27, 30, 38, 41
progenitors, viii, 7, 31, 32, 37
progeny, 3, 7, 11, 15, 16, 17
proliferation, 1, 3, 8, 11, 17, 36
property, iv, 11
prostaglandin, 13, 14, 37
prostaglandins, 14, 38
protection, vii, viii, 24, 43
protein, 13, 21, 43
proteins, viii, 19, 41, 42
protons, 7

Q

quality control, 16

R

radiation, vii, viii, 1, 2, 3, 4, 5, 6, 7, 8, 9, 10, 12, 13, 14, 16, 17, 19, 24, 26, 27, 28, 29, 30, 32, 33, 34, 35, 36, 37, 38, 39, 40, 41, 43
Radiation, 28, 29, 33, 42
radiation damage, vii, 2, 5, 7, 9, 10, 14, 16, 17, 25, 30
radio, 3, 8, 17
radioresistance, 11, 13, 15, 16, 35, 38
radiosensitization, 16, 20, 21
radiotherapy, 16, 19, 27, 40, 42
Radiotherapy, 34
random, 5, 39
range, 6, 8, 9, 16, 20, 31
rats, 15, 16, 43, 44
reaction rate, 20
reality, 23, 27
receptors, 12, 44
recovery, 32, 33, 36, 38, 44
regenerate, 1, 5
regeneration, 1, 2, 9, 10, 12, 23, 27, 28
regenerative medicine, 42
regulation, 10, 37

relationship, 6, 30, 34
relationships, 41
reoxygenation, 16, 24, 43
repair, vii, 1, 5, 7, 9, 10, 21, 27, 29, 30, 32, 44
reperfusion, vii, viii, 23, 25, 43, 44
reproduction, 1
resistance, 17, 19, 35
retinoic acid, 14, 37
Retroviral, 39
Russian, 27, 28, 29, 30, 31, 32, 33, 35, 38, 39, 40, 41, 42, 43, 44

S

scavenger, 14
search, 23
segregation, 11, 35
Self, 34
self-renewal, 2
senescence, 18, 39
Senescence, 39
sensitivity, 16, 19, 26, 28, 29, 32, 37, 38, 42
sensitization, 21
series, 7, 41
serotonin, 24, 43, 44
Sertoli cells, 17
services, iv
shape, 4, 6, 20, 33
shock, viii, 19, 41, 42, 43
short-term, 24, 41
shoulder, 5, 9, 20
side effects, 38
signaling, 11, 44
signaling pathway, 11, 44
signaling pathways, 11
signalling, 41
similarity, 7
skin, 13
small intestine, 2, 9, 10, 11, 13, 17, 23, 33, 34, 36, 40
somatic mutations, 18

somatic stem cells, 27
spermatogonial stem cells, 3, 16, 21, 39
S-phase, 13
spleen, 3, 7, 9, 20, 24, 28, 29, 30, 34, 44
squamous cell, 42, 43
squamous cell carcinoma, 42, 43
S-shaped, 5, 6
stability, 11
stabilization, 18
statistics, 5
Stem cell, 17, 27, 32, 34, 35, 37, 42
stem cells, vii, viii, 1, 2, 3, 4, 7, 8, 10, 11, 14, 15, 16, 17, 19, 23, 25, 27, 28, 29, 31, 33, 34, 35, 36, 37, 38, 39, 40, 42, 43
Stem cells, 17, 27, 32, 35, 42
strain, 12
strains, 29
strength, 27
stress, 14, 20, 44
stroma, 38, 39
stromal, 3, 28, 38
substances, 14, 16
suicide, 39
Sun, 43, 44
suppression, 14, 19, 24
survival, viii, 3, 4, 5, 6, 7, 8, 10, 12, 13, 14, 15, 17, 20, 24, 25, 28, 29, 30, 33, 34, 35, 36, 37, 38
survival rate, 33
surviving, 17
suspensions, 21
synthesis, 13, 41, 44
systems, vii, viii, 1, 2, 8, 10, 16, 17, 23, 31, 43

T

targets, vii, 5
temperature, 19, 41
tension, 44
testis, 28, 34, 40
therapeutics, 42

therapy, 8, 17, 21, 22, 32, 34, 38, 42
thyroid, 4, 29
time, viii, 6, 7, 11, 14, 15, 17, 19, 24, 25, 40
tissue, 4, 21, 25, 27, 28, 29, 40, 44
Tokyo, 29, 35
tolerance, 42
total body irradiation, 39
transcript, 32
transfection, 39
transfer, 6, 32
transgenic, 10, 34
transplantation, 4, 7, 23, 25, 43, 44
tumor, 12, 13, 19, 21, 29, 34, 35, 41, 42
tumor cells, 19, 21
tumor growth, 12
tumors, viii, 18, 38, 40
tyrosine, 32

values, 4, 6, 8, 21, 34
vector, 39
vessels, 12
villus, 13

W

Weinberg, 43
wild type, 11

X

X-axis, 5, 6, 20
X-rays, 7, 31

Y

Y-axis, 5, 20

V

validity, 24